A Historicized Ontology for the Quantum World

This book discusses the ontology in the quantum world and argues that it should be interpreted as a historicized ontology. This perspective is advocated by the quantum histories interpretation.

Inspired by the nomological interpretation from Bohmian mechanics and the corresponding understanding of the law of nature and logic, the ontology will be useful to solve the difficulties of confusing ontology and perplexing logical inference. Furthermore, this book proposes that the historicized ontology can be generalized to more complicated structures and fit naturally into temporal correlation. Core philosophical issues such as the law of nature, the ontology of events, history, and processes, and the metaphysics of history are coherently discussed, revealing the profound meaning behind the formalism of quantum histories. By providing a new framework for a historicized picture of ontology, it will effectively promote the debates in the philosophy of physics and the fundamental research in quantum mechanics.

This book will appeal to scholars and students of the philosophy of science, metaphysics, and quantum theory.

Kefu Zhu is a researcher in the philosophy of science. He obtained his PhD in the philosophy of science and technology from the University of Chinese Academy of Sciences. His main research interests encompass philosophy of physics, philosophy of logic, philosophy of artificial intelligence, and scientific metaphysics.

A Historicized Ontology for the Quantum World

A Philosophical Investigation of the Quantum Histories Interpretation

Kefu Zhu

Routledge
Taylor & Francis Group
LONDON AND NEW YORK

First published 2026
by Routledge
4 Park Square, Milton Park, Abingdon, Oxon OX14 4RN

and by Routledge
605 Third Avenue, New York, NY 10158

Routledge is an imprint of the Taylor & Francis Group, an informa business

British Library Cataloguing in Publication Data
A catalogue record for this book is available from the British Library

ISBN: 978-1-041-12070-4 (hbk)
ISBN: 978-1-041-12073-5 (pbk)
ISBN: 978-1-003-66286-0 (ebk)

DOI: 10.4324/9781003662860

Typeset in Times New Roman
by codeMantra

This book is sincerely dedicated to the four founders of the histories interpretation of quantum mechanics:

Murray Gell-Mann (1929–2019)
Roland Omnès (1931–2022)
James Burkett Hartle (1939–2023)
Robert Budington Griffiths

Contents

1 Introduction

This chapter, along with the next two chapters, sets the basic context for the readers to go into the topic of this book. Physicists, mathematicians, and philosophers struggle to clarify the foundations of quantum mechanics and debate on the interpretations of quantum mechanics. What is the foundation of quantum mechanics and how to interpret the formalism of quantum mechanics are the core issues of concern in the philosophy of physics. These two questions also involve fundamental issues in the general philosophy of science and even metaphysics and epistemology. This chapter introduces the original thoughts, historical developments, and current status of the interpretation of quantum mechanics, as well as its connection with the philosophy of science and metaphysics; it also analyzes the classifications by Bell, Leifer, and Cabello on the spectrum of interpretations of quantum mechanics, which can be regarded as a sort of "navigation map". On this basis, the main object of discussion in this book is introduced – the quantum histories interpretation.

1.1 The background of interpreting quantum mechanics

Interpretive issues in the exact sciences can be systematically categorized into two primary domains. First, from an internal theoretical perspective, key concerns include: the relationship between formalisms, meanings of symbolics, and empirical data; the ontological commitments and nomological relationships of theories (particularly in natural sciences); the deterministic versus non-deterministic nature of theories; and the causal structures embedded within theoretical frameworks. Second, from an external or inter-theoretical standpoint, critical considerations encompass the boundaries and limits of a theory, its metatheoretical properties such as consistency, reliability, and completeness, the interplay between theoretical models and mathematical structures, the equivalence/non-equivalence of different formalisms, and the reductionist or emergentist relations among theories.

Among the exact sciences, physics stands out as the most paradigmatic discipline, focusing on the universal and fundamental laws governing material changes across all spatiotemporal scales. Within this context, quantum mechanics emerges as the most contentious branch of physics, subject to a multitude of competing interpretations – akin to being "surrounded by a swarm of bees". Since its inception

DOI: 10.4324/9781003662860-1

in the 1920s, quantum mechanics has been extensively and profoundly applied across various subfields of physics, as well as in chemistry, biology, astronomy, materials science, computer science, medicine, and numerous other disciplines. Through the relentless efforts of pioneering figures such as Bohr, Einstein, Dirac, Heisenberg, von Neumann, Weyl, Wigner, Born, Jordan, and Feynman, the mathematical formalism of quantum mechanics has been rigorously developed and refined, achieving high-precision validation through laboratory experiments and cosmological observations. Despite these advancements, significant disagreements persist regarding the essential nature and interpretation of quantum mechanics.[1] The investigation of its foundational principles by physicists, mathematicians, and philosophers remains deeply intertwined with ongoing debates over its interpretive frameworks.

As Laloë has pointed out,

> It has often been claimed, including especially by the founding fathers of quantum mechanics, that this theory has a significant impact on the philosophical view we have of the universe. In retrospect, one century later, this strong impact of physics on philosophy seems to have been overestimated: various interpretations have since been proposed, supporting different philosophical views, and still no definitive logical argument has been found to select one or the other.
>
> (Laloë, 2022: 45)

It is not an issue that can be definitively resolved which philosophical positions quantum mechanics supports or refutes. Philosophical perspectives that were initially fiercely challenged by an early understanding of quantum mechanics may, after many years, have been revived, while viewpoints once deemed convincing have subsequently been subject to comprehensive scrutiny and gradually discarded. Consequently, when reviewing the century-long history of quantum mechanics interpretation, it becomes evident that the entanglement between quantum mechanics and its philosophical interpretations is highly complex and intertwined. Physics does not merely deliver verdicts on philosophical positions, nor can any existing philosophical perspectives be straightforwardly applied to the understanding of physical phenomena and physical theories.

As one of the foundational pillars of modern physics, what constitutes the underlying foundations of quantum mechanics? How should its abstract mathematical framework be interpreted? These two questions represent the core issues that the philosophy of physics, as a central branch of the philosophy of science, seeks to address. Simultaneously, they also intersect with numerous topics in general philosophy of science, metaphysics, and epistemology. Researchers aim to comprehend the essence of the fundamental concepts and the formalism of quantum mechanics, explain phenomena at the quantum scale that differ from those observed at the classical scale, and clarify the relationship between reality, laws, representation, and observation within the world picture of quantum mechanics. More importantly, this work serves as a necessary conceptual foundation for the development of quantum gravity or unified theories. The following list of questions elaborates on the aforementioned core issues:

- Quantum mechanics employs precise mathematical languages and structures (formalism), but is interpretation an inherent requirement of studying quantum mechanics? Do any so-called interpretations genuinely enhance our understanding of quantum mechanics? In practice, many people prioritize the predictive power of quantum mechanics, which enables its practical application (Drummond, 2019: 391). These questions themselves scrutinize the rationality of the research direction concerning "interpretations of quantum mechanics".
- Within the formalism of quantum mechanics, which components can be considered real and possess ontological status? If the quantum state is deemed real, how can its characteristics, which diverge from classical physical reality, be explained?
- What physical mechanism underlies the measurement process in quantum mechanics? Does quantum mechanics necessarily require the introduction of an external observer when describing the measurement process? Does quantum measurement demand additional principles beyond the basic tenets of quantum mechanics to achieve clarity? Is there a measurement process that fundamentally differs from the unitary evolution of quantum systems?
- Given that quantum mechanics is a probabilistic theory, is it inherently deterministic or non-deterministic? Does traditional causality still apply in this context? Does this necessitate a reevaluation of the interplay between probability, causality, and determinism?
- The existence of various paradoxes in quantum mechanics suggests that classical physical laws or logical rules we take for granted may need revision or abandonment. Can these paradoxes be resolved through logical or physical reasoning?
- How should time be understood within the framework of quantum mechanics? What distinguishes the quantum state at a single moment from a sequence of quantum events involving temporal progression?
- Is quantum mechanics a complete fundamental theory or merely an incomplete effective theory? If the latter, what essential features should the underlying theory encompass?
- Can quantum mechanics accurately describe all macroscopic and microscopic phenomena? Can it adequately account for the evolution of the entire universe? If not, modifications to quantum mechanics would be required. What experimental observations could distinguish such a modified theory from the current formulation of quantum mechanics?
- How should locality, causality, and contextuality be understood in quantum mechanics? Which holds greater fundamental significance – quantumness or classicality (characterized by locality, determinism, causality, and non-contextuality)? Should these two coexist and complement each other in physical descriptions, as suggested by the Copenhagen interpretation? Can quantum physics be reduced to a theory embodying classicality, or is the world fundamentally

quantum, with classicality emerging as an approximation of the quantum realm? If the quantum world is fundamental, how does classicality emerge from quantumness?

Additionally, questions arise regarding the necessity of distinguishing between observers and the observed, the boundary between the macroscopic classical world and the microscopic quantum world, and the compatibility and integration of the conceptual frameworks of quantum mechanics with other fundamental theories, such as general relativity and thermodynamics.

These questions transcend mere technical concerns in physics, delving into philosophical and conceptual domains. The philosophy of physics not only strives for clearer and more accurate understandings of fundamental physical concepts but also examines the compatibility among the foundational principles of a theory and the compatibility between different theories. This involves multiple aspects, including the ontological assumptions of theories, relationships between theoretical entities, and logical consistency both within and across theories.

Despite the persistent efforts of numerous philosophers of physics and physicists to seek reasonable answers to these questions, progress has been limited. In his article "Interpretations of Quantum Theory: A Map of Madness", quantum information scientist Adán Cabello fictionalizes a dialogue between a journalist and physicists, highlighting significant disagreements in their perspectives:

> Motivated by some recent news, a journalist asks a group of physicists: "What's the meaning of the violation of Bell's inequality?" One physicist answers: "It means that non-locality is an established fact". Another says: "There is no non-locality; the message is that measurement outcomes are irreducibly random". A third one says: "It cannot be answered simply on purely physical grounds, the answer requires an act of metaphysical judgement". Puzzled by the answers, the journalist keeps asking questions about quantum theory: "What is teleported in quantum teleportation?" "How does a quantum computer really work?" Shockingly, for each of these questions, the journalist obtains a variety of answers which, in many cases, are mutually exclusive.
>
> (Cabello, 2017: 138)

This dialogue underscores the lack of consensus regarding the interpretation and understanding of quantum mechanics, resulting in no unified or consistent answers to these seemingly fundamental questions. Such ambiguities may persist indefinitely. While many physicists disregard these topics during specific research endeavors or actively avoid them, they inevitably return to these foundational interpretational issues when addressing cutting-edge research that requires a deeper exploration of basic concepts, reflection on theoretical frameworks, and integration of the underlying logic across disciplines.

A brief review of the historical development of quantum mechanics interpretations reveals a rich tapestry of approaches. Shortly after Schrödinger established wave mechanics, early interpretations emerged, including Schrödinger's "wave realism" (Schrödinger, 1927a,1927b, 1927c), Madelung's fluid mechanics analogy (Madelung, 1927), and de Broglie's pilot wave theory (de Broglie, 1927). Subsequently, the Copenhagen interpretation, synthesizing insights from Bohr, Born, Heisenberg, von Neumann, and others, gained widespread acceptance alongside Einstein's ensemble interpretation. Mid-20th century developments introduced Bohmian mechanics (Bohm, 1952; also see Bub, 2022), Everett and DeWitt's relative state/many-worlds interpretation (DeWitt and Graham, 1973), and the many-minds interpretation (Albert, 1992; Stapp, 2004). By the 1980s, alternative interpretations modifying the quantum mechanics framework appeared, such as spontaneous collapse theories (Ghirardi, Rimini and Weber, 1986). See also Jammer (1989), Laloë (2019), Greenberger, Hentschel and Weinert (2009), and Olival Freire (2022). Each interpretation offers unique starting points and philosophical stances, demonstrating explanatory power for certain problems while encountering insurmountable challenges. Physicist David Mermin's humorous remark, "New interpretations appear every year. None ever disappear" (Mermin, 2012: 8), reflects both the ongoing pursuit of satisfactory interpretations and the widening divergence of perspectives.

The subsequent discussion will examine the quantum histories interpretation, a representative approach to addressing these interpretational issues. Proposed by Griffiths et al., this interpretation posits "quantum histories" as its fundamental concept. Without altering the basic structure of quantum mechanics, it aims to resolve difficulties and paradoxes encountered by prior interpretations in a consistent and reasonable way. By treating all physical processes in the universe as fundamentally quantum mechanical, it explains the measurement process without invoking external observers. Furthermore, it seeks to apply its formalism to arbitrary closed systems, including the universe as a whole, deriving mechanisms by which classicality emerges from quantumness. If finally proven more reasonable and consistent by more and more theoretical and empirical evidence, the quantum histories interpretation could provide valuable insights into the foundation of quantum mechanics.[2]

1.2 Status quo of the pursuit of a consistent and reasonable interpretation of quantum mechanics

To date, scholars have proposed numerous interpretations of quantum mechanics from diverse perspectives. Despite these efforts, the physics and philosophy communities continue to debate and dispute this topic, as no single interpretation fully satisfies all criteria or addresses every aspect comprehensively. Many interpretations exhibit trade-offs, prioritizing solutions to specific challenges while exposing weaknesses in other areas, thereby incurring significant theoretical costs. Currently, researchers not only seek the most accurate and reasonable interpretation but also aim to identify commonalities and distinctions among various interpretations during this exploration. Additionally, there is interest in determining whether interpretations can be appropriately categorized based on their essential features. Some scholars have proposed classification standards for this purpose. For example, John Bell introduced a framework that divided the six most prominent interpretations of his time into two categories: "romantic" and "unromantic" (Bell, 1987: 181–195). The "romantic" category includes the Copenhagen interpretation, the von Neumann–Wigner collapse interpretation, and Everett's relative state/many-worlds interpretation, whereas the "unromantic" category encompasses the pure pragmatic view, the stochastic process interpretation, and de Broglie–Bohm's pilot wave theory.

If we adopt the realism/anti-realism stance frequently discussed in the philosophy of science, various interpretations can be categorized into three primary groups: anti-realism, moderate realism, and full-blooded realism (Leifer, 2005). Anti-realism includes instrumentalism, operationalism, and the Copenhagen orthodox interpretation. Moderate realism encompasses quantum logic realism, Everett's theory, and the quantum histories theory. Full-blooded realism, favored by Bell, includes Bohmian-type theories and spontaneous collapse theories.

In the previously cited article, Cabello provides an alternative classification (Cabello, 2017: 139), namely intrinsic realism and participatory realism. The first category posits that the probability distribution of quantum measurement results is determined by the intrinsic properties of the observed system. This can be further subdivided into two subcategories: the "ψ-ontic" view, which regards the quantum state as the intrinsic property of the observed system, and the "ψ-epistemic" view, which considers the quantum state as representing knowledge of the underlying objective reality – similar to how classical statistical mechanics assigns a probability distribution to each point in phase space. The second category, participatory realism, argues that the probability distribution of quantum measurement results is not determined solely by the intrinsic properties of the observed system. While it does not deny the existence of an objective world, quantum theory focuses on the experiences or interactions of the observer or agent with the observed system rather than directly addressing the intrinsic properties of the system.[3] Examples from both categories are as follows:

Intrinsic realism:

1 ψ-ontic: Bohm mechanics, many-worlds interpretation, modal interpretations, Bell's "beables", and collapse theories (which modify quantum mechanics and thus extend beyond mere interpretation).
2 ψ-epistemic: Einstein's perspective, Ballentine's approach, quantum histories, and Spekkens's framework.

Participatory realism:

1 Knowledge-based: Copenhagen interpretation, Wheeler's views, relational interpretations, Zeilinger's stance, the "no interpretation" perspective, and Brukner's contributions.
2 Belief-based: Quantum Bayesianism (which could also be abbreviated as QBism).

These classifications provide a broad overview of the interpretational landscape of quantum mechanics. It becomes evident that different philosophical positions lead to significantly divergent understandings of the fundamental picture of quantum mechanics. To date, extensive literature has been accumulated for each of the interpretations mentioned above, including original expositions by their founders, as well as discussions, critiques, and clarifications regarding specific issues. Here, we only outline the general situation without delving into detailed elaborations.

The quantum histories interpretation, which this book focuses on, represents a relatively recent development, first proposed by Griffiths in 1984. However, this interpretation is not without precedent; its intellectual lineage can be traced back to earlier frameworks such as the many-worlds interpretation, Feynman's sum-over-histories concept, and decoherence theory (see Chapter 3).

From the existing literature, it is clear that research on quantum histories interpretation has become highly developed since its introduction by Griffiths (1984). Discussions have emerged from various perspectives, and the literature can be broadly classified into several types:

1 Original papers by the main founders – Griffiths, Omnès, and Gell-Mann and Hartle – along with their elaborations and extensions of their respective viewpoints. (Among the latter two, Gell-Mann was Hartle's doctoral supervisor, and most of their related papers were co-authored.)
2 Critiques and corrections of this interpretation by scholars from diverse standpoints, including supporters of Bohmian mechanics, spontaneous collapse theory, and relational interpretations, as well as rebuttals and debates by the founders themselves.
3 Applications of this interpretation to analyze additional observational phenomena, along with attempts to supplement, refine, and develop it further.

   Other literature also explores:

   - Using this interpretation to address and resolve various quantum paradoxes.
   - Comparing and analyzing this interpretation alongside other significant interpretations of quantum mechanics.
   - Applying the formalism of quantum history to discuss classical field theory, relativistic quantum mechanics, quantum field theory, quantum cosmology, and quantum information science.
   - Modifying and expanding the formalism through research on discrimination conditions for other history families, attempts to eliminate the decoherence functional, defining generalized probability assignment methods for quantum history, and employing advanced mathematical tools to reformulate the quantum history framework.

Notes

* In order to enhance the fluency and professionalism of the English in the draft, the Youdao AI Translation Polishing Tool (March 2025 version) was employed in part of the original manuscript.

1 It is said that the renowned physicist Feynman once quipped: "I think I can safely say that nobody really understands quantum mechanics".

2 This book primarily focuses on the interpretation of non-relativistic quantum mechanics. Given that relativistic quantum mechanics, quantum statistical physics, and quantum field theory involve the quantization of infinite-dimensional systems, their theoretical frameworks are significantly more intricate, and their mathematical formalisms remain insufficiently rigorous. These factors introduce additional challenges in their interpretation, so I have to defer discussion of these topics for the present. Nevertheless, even within the relatively well-established domain of non-relativistic quantum mechanics, numerous challenging interpretive issues have arisen as understanding has deepened.

3 One can refer to a famous saying by Niels Bohr: "There is no quantum world. There is only an abstract quantum physical description. It is wrong to think that the task of physics is to find out how nature is. Physics concerns what we can say about nature" (Petersen, 1963: 12).

2 The formalisms and interpretations of quantum mechanics

This chapter, based on the previous introduction of the background information, reviews the standard formalism of quantum mechanics – its fundamental mathematical structures (the basic contents of Hilbert space representation and operator algebra) and physical assumptions (kinematics, dynamics, and the quantum measurement theory). These are the premise and foundation for discussing the quantum histories interpretation. Then, in combination with the viewpoints of some literature in recent years, we explore what counts as the interpretation of quantum mechanics, analyze the relationship between the theory of quantum mechanics and its interpretation, and declare the author's basic stance on these two issues. These two are the primary issues that all further interpretive work in the philosophy of physics has to face (not limited to the interpretation of quantum mechanics), and the necessary clarification here is also conducive to the subsequent discussion. Three widely used terms – theory, formalism, and interpretation – are explicated and differentiated by offering the necessary definitions of their meanings, based on the works of Acuña and Ruetsche. The clarification will help to make the related discussions more precise.

2.1 Formalism and the fundamental axioms of quantum mechanics

First, we briefly review the formalism of standard quantum mechanics, especially its mathematical structure and fundamental physical assumptions. This part of the contents can be found in the common presentations of textbooks on quantum mechanics, mathematical physics, and functional analysis. Hence, we will not specifically indicate the original sources (see Dirac, 1930; Von Neumann, 1932; and recently published excellent introductions, for example, David, 2015; Landsman, 2017).

The widely recognized mathematical representation of standard quantum mechanics in textbooks is realized through Hilbert space and the operators acting on it. A Hilbert space refers to a complete vector space with an inner product

DOI: 10.4324/9781003662860-2

operation and a metric induced by the inner product (thereafter, only complex Hilbert spaces are considered), defined by the following axioms:

1 A complex vector space $\mathcal{H}$;
2 A binary operation on it, $\langle x, y\rangle : \mathcal{H} \times \mathcal{H} \mapsto \mathbb{C}$, $\forall x, y \in \mathcal{H}$, which is called the inner product, satisfying:

    1 Conjugate symmetry, $\langle x, y\rangle = \langle y, x\rangle$;
    2 Linearity with respect to the first variable, $\langle ax_1 + bx_2, y\rangle = a\langle x_1, y\rangle + \langle bx_2, y\rangle$, $\forall a, b \in \mathbb{C}$, $\forall x_1, x_2, y \in \mathcal{H}$.

       From (1) and (2), we can see that the inner product is "conjugate linear" with respect to the second variable;

    3 Positive definiteness, $\langle x, x\rangle > 0$, if $x \neq 0$; $\langle x, x\rangle = 0$, if $x = 0$.

       From this requirement, the metric in the Hilbert space can be defined as

$$d(x, y) = \|x - y\| = \sqrt{\langle x - y, x - y\rangle}$$

       and the norm is defined as

$$\|x\| = \sqrt{\langle x, x\rangle}$$

Thus, it also constitutes a metric space. Besides, the inner product operation also requires completeness, that is, every Cauchy sequence converging under this norm in the space must converge to some element in the space.

The vectors and dual vectors in it can be intuitively denoted via Dirac's notation, that is, each vector in the Hilbert space can be written as $|\psi\rangle$, and its dual vector (a mapping from the vectors in the Hilbert space to the complex number field) can be correspondingly written as $\langle\psi| := \langle\psi, \cdots\rangle$. Each vector physically represents a pure quantum state (but mixed states can only be represented by density matrices, see below).

A (bounded linear) operator on the Hilbert space $\mathcal{H}$ refers to a linear mapping $\hat{O} : \mathcal{H} \mapsto \mathcal{H}$, which must satisfy:

$$\hat{O}(\alpha\psi + \beta\phi) = \alpha\hat{O}(\psi) + \beta\hat{O}(\phi)$$

and

$$\|\hat{O}\| = \sup_{\{\|x\|=1\}} \|\hat{O}x\| < \infty$$

If there exist $\psi \in \mathcal{H}$ and $k \in \mathbb{C}$ such that

$$\hat{O}\psi = k\psi$$

then ψ is called an eigenvector of $\hat{O}$ and k is the corresponding eigenvalue of $\hat{O}$ for this eigenvector ψ. Only eigenvalues are the data directly obtained from experiments, while states and vectors can only be inferred indirectly. In physics, the result data of experimental observations must be real numbers, so generally only Hermitian operators (in the mathematical literature, they are called self-adjoint operators) are accepted for constructing a physical theory, that is, it is required that

$$\langle \hat{O}\psi, \phi \rangle = \langle \psi, \hat{O}\phi \rangle \quad \forall \psi, \phi \in \mathcal{H}$$

The eigenvalues of such operators are all real, and there exists a decomposition of $\mathcal{H}$ with the eigenvectors of this operator as an orthogonal basis.

If ψ and ϕ are eigenvectors of $\hat{O}$ corresponding to the same eigenvalue, then $\alpha\psi + \beta\phi$ ($\alpha, \beta \in \mathbb{C}$) is also an eigenvector corresponding to this eigenvalue. The trace of an operator $Tr(\cdots)$ can be used to obtain the sum of all its eigenvalues.

$$Tr(\hat{O}) = \sum_n k_n$$

The operation represented by the projection operator (projector) $\hat{P}_j$ is to project onto a subspace $\mathcal{J} \subseteq \mathcal{H}$ of the Hilbert space, defined as follows:

$$\hat{P}_j(\phi) = \begin{cases} \phi, \text{ if } \phi \in J \\ 0, \text{ if } \phi \text{ is orthogonal to } J \end{cases}$$

Projectors possess the following important properties:

$$\hat{P}_j^2 = \hat{P}_j \hat{P}_j^\dagger$$

Hermitian operators are used to represent observable physical quantities (they will be referred to as "observables" for short): each eigenvalue (i.e., the "spectrum" of the operator) represents a possible observed value, and the vector that only yields the value q after the corresponding (to this operator) measurement is the eigenvector of the operator with the eigenvalue q. Taking the projector as an example, $\hat{P}_j$ represents a physical quantity: for all states $\phi \in \mathcal{J}$, the observed value is 1 after the measurement; while for all those states ϕ orthogonal to $\mathcal{J}$, the observed value is 0 after the measurement.

For a given operator and its eigenvector ϕ_q with the eigenvalue q (without considering degeneracy), the probability of obtaining the result q when implementing observation with this operator in any pure state ψ can be calculated:

$$P = \left| \langle \psi, \phi_q \rangle \right|^2$$

The expectation value of the measurement of the operator $\hat{O}$ on any pure state ψ is

$$\langle \psi, \hat{O}\psi \rangle$$

The definition of the density matrix is: $\rho := \sum_i p_i |\psi_i\rangle\langle\psi_i|$, where $p_i > 0$ and $\sum_i p_i = 1$. Here, ψ_i represents some pure state, and p_i represents the probability of occurrence of the corresponding pure states. When $i = 1$, ρ is the density matrix of a pure state; when $i > 1$, ρ is the density matrix of a mixed state. The density matrix itself is also an operator. The density matrix of a pure state satisfies the following properties:

$$\langle \psi, \rho\psi \rangle > 1, \text{ for any } \psi \in \mathcal{H}$$

$$\rho^2 = \rho$$

$$\rho^\dagger = \rho, \; \rho^\dagger \text{ denotes its transposed conjugate.}$$

$$Tr(\rho) = 1$$

The average value obtained by applying an operator to ρ can be concisely expressed as: $\langle \hat{O} \rangle = Tr(\hat{O}\rho)$.

Apart from studying the various properties of states on Hilbert space, we can also focus solely on the various composite operations between operators on Hilbert space and the algebraic structures they form. This is the algebraic perspective of quantum mechanics, concerning the subject of operator algebra. A *-algebra $\mathfrak{U}$ refers to a complex vector space that possesses the following operations:

1 A binary operation, multiplication: for $X, Y \in \mathfrak{U}$, $XY \in \mathfrak{U}$;
2 An identity element: $\mathbf{1} \in \mathfrak{U}$;
3 A unary operation, involution, denoted by asterisk *: for $X \in \mathfrak{U}$, $X^* \in \mathfrak{U}$.

These three operations satisfy the following conditions for any $X, Y \in \mathfrak{U}$ and $\alpha \in \mathbb{C}$:

$$(XY)Z = X(YZ)$$

$$X\mathbf{1} = X = \mathbf{1}X$$

$$X(Y+Z) = XY + XZ$$

$$(X+Y)Z = XZ + YZ$$

$$X^{**} = X$$

$$(X+Y)^* = (X^* + Y^*)$$

$$(\alpha X)^* = \bar{\alpha} X^*$$

$$(XY)^* = Y^* X^*$$

$$\mathbf{1}^* = \mathbf{1}$$

To define a C^*-algebra over a *-algebra, it is also required that the norm of operator $\|\cdots\| : \mathfrak{U} \mapsto \mathbb{R}^{\geq 0}$ forms a complete topological vector space, and in the meantime, the norm and the involution must satisfy:

$$\|XY\| \leq \|X\| \, \|Y\|$$

$$\left\|X^* X\right\| = \left\|X\right\|^2$$

From this, it can be seen that Hermitian operators are the self-adjoint operators here: $X^* = X$. From the algebraic perspective, the "state" defined earlier is a positive definite, unit-preserving linear functional $\omega : \mathfrak{U} \mapsto \mathbb{C}$ on the C^*-algebra, that is,

$$\omega\left(X^* X\right) \geq 0$$

$$\omega\left(\mathbf{1}\right) = \mathbf{1}$$

The set of all operators on a Hilbert space $\mathcal{H}$, $B\left(\mathcal{H}\right)$, forms a C^*-algebra, where the unit element is the identity operator, the product is the composition of operators, and the involution * maps an operator to its adjoint operator:

$$\left\langle \hat{O}\psi, \phi \right\rangle = \left\langle \psi, \hat{O}^* \phi \right\rangle$$

The state ω on $B\left(\mathcal{H}\right)$ can be defined using the density operator ρ on this Hilbert space as

$$\omega\left(\hat{O}\right) = Tr\left(\rho \hat{O}\right)$$

In quantum mechanics, the C^*-algebra of operators is generally a non-commutative algebra. The following theorem points out a fundamental connection between *-algebras and Hilbert spaces:

Any finite-dimensional *-algebra can be represented as a subalgebra of $B\left(\mathcal{H}\right)$ on some Hilbert space $\mathcal{H}$.

It is worth noting that even in the relatively simple case of non-relativistic quantum mechanics, in addition to the Hilbert space formulation and the operator algebra formulation mentioned above, there are also diverse sorts and varieties of formulations based on other mathematical structures. Mathematicians and physicians with different interests and from different research branches have constructed what they consider to be a convenient and accurate axiomatization of quantum mechanics, and at the same time, they also need to ensure that this formalism can be reduced in a way back to the usual Hilbert space representation under certain conditions. Some tentative programs include formalizing quantum mechanics on a complex geometric structure (Brody and Hughston 2001; Giachetta, Mangiarotti and Sardanashvily, 2010; Roberts and Teh, 2016; Ciaglia, Ibort and Marmo, 2019a), based on topos theory (Döring and Isham, 2010; Flori, 2013; Flori, 2018), based on groupoid theory (Ciaglia, Ibort and Marmo, 2019b; Ciaglia et al., 2020), through quantum set theory formulation (Takeuti, 1981; Ozawa, 2016; Ozawa, 2021), and so on. Some studies point out that some of these programs can actually be unified or mutually implied (for example, Döring, Eva and Ozawa, 2021). But generally, most of these attempts have only just begun with preliminary axiomatization and simple analysis of a few cases. Further follow-up and credible handling of various quantum paradoxes are expected but still inadequate. The complicated and abstract mathematical structures and their connection with physical reality call for more interpretative work than ever before.

The development of quantum histories interpretation also requires higher mathematical tools. As for which mathematical constructions can naturally and appropriately reflect the essence of quantum mechanics and be coherent with other fundamental physical theories, more attempts are needed. The histories approach to non-relativistic quantum mechanics in existing literature is mainly limited to the use of Hilbert space and its operators. When it comes to quantum field theory, tools such as path integrals and C^*-algebras are used. Limited by space, this will not be discussed further here.

The following are the basic physical assumptions of quantum mechanics:

First, the physical state of a microscopic particle can be characterized by a quantum state, and its motion can be described by a wave function $\psi(\vec{x},t)$ with space-time coordinates as independent variables.

Second, for a quantum system, if it may be in states $\psi_1,\ldots,\psi_n$, then the linear superposition of these states is also a possible state of the system:

$$\Psi = \sum_i c_i \psi_i$$

where c_i is the coefficient for the superposition.

Third, the evolution of the state vector ψ representing the state of an isolated quantum system over time follows the Schrödinger equation:

$$i\hbar \frac{\partial \psi}{\partial t} = \hat{H}\psi$$

Fourth, each dynamical quantity O in quantum mechanics is represented by a linear Hermitian operator $\hat{O}$.

Fifth, when a dynamical quantity O is observed, the observed value can only be one of the eigenvalues of the corresponding linear Hermitian operator $\hat{O}$. If the system is in the state

$$\Psi = \sum_i c_i \psi_i$$

and $\{\psi_i\}$ is the complete orthonormal basis of eigenfunctions of $\hat{O}$, $c_i = \langle \psi_n | \Psi \rangle$ is the corresponding expansion coefficient, then the probability of measuring the eigenvalue O_i is $|c_i|^2$.

From these assumptions, the basics of kinematics, dynamics, and measurement theory of quantum mechanics can be derived.

2.1.1 Kinematics

In non-relativistic quantum mechanics, the most fundamental kinematic variables are the generalized coordinates $\hat{q}$ and generalized momenta $\hat{p}$. The commutator $\left[\hat{a},\hat{b}\right]$ between operators $\hat{a}$ and $\hat{b}$ is defined as

$$\left[\hat{a},\hat{b}\right]=\hat{a}\hat{b}-\hat{b}\hat{a}$$

Unlike the Poisson bracket $\left\{q_i,p_i\right\}$ satisfied by the coordinates and momenta in classical mechanics, this commutator satisfies the Heisenberg algebra with the anti-commutation relation:

$$\left[\hat{q}_i,\hat{p}_i\right]=i\hbar\hat{\delta}_{ij},\text{ where }\hat{\delta}_{ij}=\begin{cases}0,\text{ when } i\neq j\\1,\text{ when } i=j\end{cases}$$

$$\left[\hat{q}_i,\hat{q}_i\right]=\left[\hat{p}_i,\hat{p}_i\right]=0$$

The first equation shows the essential difference between the quantum mechanical coordinate-momentum commutation relation and the Poisson bracket $\left\{q_i,p_i\right\}$ in classical mechanics: coordinates and momenta do not commute. In the quantum histories interpretation, the non-commutativity is the key for determining whether different history families are compatible or not. When explaining quantum mechanical paradoxes with temporal event sequences in quantum histories, a careful analysis of the unnoticed non-commutativity is indispensable.

2.1.2 Dynamics

Dynamics provides the evolution law of the system over time. In quantum mechanics, dynamics is reflected through the time evolution of the state wave function (Schrödinger picture) or the time evolution of the observable operator (Heisenberg picture). The forms of the dynamical equations in the two pictures seem different, but they are essentially equivalent. Given the Hamiltonian of the form

$$\hat{H}=\frac{\hat{p}^2}{m}+V\left(\vec{r},t\right)$$

The Heisenberg equations for the coordinate and momentum operators are

$$\frac{d\hat{q}_i}{dt}=\frac{1}{i\hbar}\left[\hat{q}_i,\hat{H}\right]$$

$$\frac{d\hat{p}_i}{dt}=\frac{1}{i\hbar}\left[\hat{p}_i,\hat{H}\right]$$

For general dynamical quantities, the equation is

$$\frac{d\hat{O}_i(q,p)}{dt} = -\frac{i}{\hbar}\left[\hat{O}_i, \hat{H}\right]$$

The evolution equation of the quantum state is the Schrödinger equation:

$$i\hbar\frac{\partial\psi}{\partial t} = \hat{H}\psi$$

It can be proved that the Schrödinger equation and the Heisenberg equation are mathematically equivalent, and they can give the same results when calculating the expectation values of observables (this stems from the equivalence of matrix mechanics and wave mechanics). Particularly, if represented by density matrix $\rho = |\psi\rangle\langle\psi|$, the dynamical equation in the Heisenberg picture can be written as

$$\frac{d\rho(t)}{dt} = \frac{1}{i\hbar}\left[\hat{H}, \rho(t)\right]$$

2.1.3 *Quantum measurement theory*

Unlike the cases in classical mechanics, quantum mechanics must address the indirect correspondence between the theoretical object and the measurement process, and answer the question of how to obtain empirical measurement values from abstract dynamical quantity operators and quantum states through experiments and calculations. The measurement theory of quantum mechanics includes the following basic rules:

1 What the wave function represents is the probability amplitude of the state, not the probability itself. The square of its norm represents the probability density function of the particle being in a certain physical state.
2 Each time a mechanical quantity $\hat{O}$ is measured, only its eigenvalue $\hat{O}_n$ can be obtained, with a probability of p_n corresponding to that eigenvalue. The expectation value obtained from multiple measurements is

$$\langle\psi|\hat{O}|\psi\rangle = \sum p_n O_n$$

After expanding the wave function in terms of its eigenstates, the square of the norm of the coefficient in front of each term is the probability of observing that eigenvalue in that state:

$$\psi = \sum C_n \psi_n$$

$$p_n = |C_n|^2$$

For a general quantum measurement, if a set of operators $\{\hat{M}_i\}$ is used to describe it, and these operators satisfy the completeness condition:

$$\sum_m \hat{M}_m^\dagger \hat{M}_m = \hat{I}$$

where $\{m\}$ represents all possible measurement outcomes, then the probability of obtaining the measurement result m is

$$p(m) = \langle \psi | \hat{M}_m^\dagger \hat{M}_m | \psi \rangle$$

After the measurement, the state of the system becomes

$$\psi_m = \frac{\hat{M} | \psi \rangle}{\sqrt{\langle \psi | \hat{M}_m^\dagger \hat{M}_m | \psi \rangle}}$$

2.2 The standards of interpreting the formalism of quantum mechanics

After introducing the core definitions, mathematical structures, and physical assumptions of quantum mechanics, the next question arises: Is the basic framework of quantum mechanics sufficient to explain all the experimental results of quantum mechanics and have the necessary predictive power? If the answer is affirmative, then why is there any need for interpretation? What are the differences and similarities between "theory", "formalism", and "interpretation"? This should be clarified before any discussion on the interpretation issue is carried out.

From the classification of various interpretations mentioned in Chapter 1, it can be seen that different philosophical positions will lead to very different understandings of the overall picture. Then, does the content provided by standard quantum mechanics count as a complete theoretical picture? Why do such different understandings arise? What exactly does "interpreting it" mean and what is the goal of interpretation?

Unfortunately, many studies fail to clearly define at the beginning the connotations of the terms such as "interpretation", "theory", "system", and "approach" they use, which leads to a spread of confusion. Some people have already realized the disorderly use of terms. Laloë F. believes that standard quantum mechanics is actually a general methodology, a framework that can develop multiple theories (Laloë, 2019: xiii); Dieks (2017: 295–314) also believes that various interpretations of quantum mechanics are actually rival theories or many "different quantum theories" (different quantum theories). The following will refer to Acuña (2019)'s review to introduce the analyses of several physical philosophers in recent years, specifically on what is an interpretation of a physical theory, and thereby make some necessary clarifications.

Acuña reviews the representative opinions of three philosophers, pointing out that they are actually very similar, and then presents his own view.

The first one is Ruetsche (2011: 8–9). She believes that to interpret a physical theory, the first stage to do is to "summarize what kind of structure the theory uses to represent physical reality". This stage can be divided into three steps:

The first step is to describe the state space of the theory;

The second step is to provide the observables of the theory, and these two constitute the kinematic part of the theory (in Ruetsche's terms);

The third step is to present the dynamics part of the theory, that is, the description of the time evolution of the states and observables in it.

Take classical mechanics as an example, these three steps correspond respectively to:

1 The set of all coordinate-momentum ordered pairs (p,q) in the phase space can represent the state space, and each ordered pair (p,q) represents a state of the system;
2 The observables are various differentiable functions mapping from the state space to the set of real numbers;
3 The trajectories in the phase space determined by the Hamiltonian function and the Hamiltonian equation specify the specific dynamical evolution.

The second stage of interpretation is to expound the semantic part, that is, to determine which models are satisfied by the theoretical structure, or to say, to give which models represent the possible worlds that the theory considers physically possible. If we only consider the so-called "pristine scenario", then these physical models are directly determined by the "nomic" content in that part of the structure, that is, the laws in the structure directly determine the physical models without the need for any additional metaphysical, epistemological assumptions and other additional conceptual resources. At this time, the structure directly provides an interpretation of itself. Once the semantic part is completed, the theory can be regarded as having been interpreted.

From the above analysis, Ruetsche proposed a test (Ruetsche's test) to examine whether an interpretation work can be regarded as completed: for the system represented by the state in the theory, can the theoretical structures explain which propositions, among all the propositions that assign values to physical quantities, are the true propositions about this system? (Ruetsche, 2011: 9)

Ruetsche claims that the theoretical structure of non-relativistic quantum mechanics is already very clear, and its core parts are:

1 The observable quantities of physical systems correspond to Hermitian operators acting on the Hilbert space $\mathcal{H}$;
2 The state W of the physical system is in one-to-one correspondence with the density operator ρ on the Hilbert space $\mathcal{H}$, and the expectation value of the observable quantity A is $Tr(WA)$;

3 For a system with the initial state W_0 and the Hamiltonian H, its dynamic evolution is

$$W_t = e^{-\frac{i}{\hbar}Ht} W_0 e^{\frac{i}{\hbar}Ht}.$$

Although these three parts have "completed" the kinematics and dynamics of quantum mechanics, these "nomic" contents cannot yet determine their model, so the pristine picture has not yet arisen. People can calculate the probability distribution of different evolutions of the system by its Schrödinger equation and the given initial state information, but this only explains the probabilistic observation results. From Ruetsche's test, its semantics part is still incomplete and cannot explain which propositions that assign values to the observables of quantum mechanics (especially the non-commutative quantities) are true propositions about the system. It must rely on some external concepts to determine this. In other words, the theoretical structure of quantum mechanics cannot determine its interpretation and is insufficient to delineate which possible worlds are the "physically admissible possible worlds" of this theory.

There are many strategies to supply new conceptual resources to complete the interpretation of quantum mechanics – this is one of the major reasons for the blooming of so many competing interpretations – such as considering whether to add more underlying variables to the state of the system (hidden variables approach), whether to introduce the dynamic mechanism of spontaneous collapse of the wave packet (spontaneous collapse theory), whether quantum systems really have values that cannot be accurately predicted from their quantum states, and whether quantum reality leads to one or more worlds (many-worlds theory). It can be said that adding these conceptual resources aims to complete a series of more complete (in the sense of the completeness of formal theories) new theories from the existing standard quantum mechanics.[1]

The second view comes from the Dutch philosopher Muller (2015). He calls the core of the standard quantum mechanics formalism "minimal quantum mechanics" (denoted as QM_0) – in his opinion, this is the simplest formalism that can "calculate the probability distribution of quantum measurement results". The core selected by Muller includes: the connection between the Hilbert space with its states or density operators and the states of the quantum system at any time, the evolution of the state over time following the Schrödinger equation, the observables are represented by Hermitian operators in the Hilbert space, and the formula for solving the expectation value of the measurement results of the observables.

Obviously, the QM_0 selected in this way still has many deficiencies. For example, its description of the meaning of "state" is also empty; the state is only regarded as a mathematical tool for calculating probabilities. Moreover, because it cannot explain the phenomenon of superposition states, the observables here cannot even be understood as "attributes", at most, they can only be regarded as a calculation tool. Muller points out that such a "minimal" version of QM_0 with the most basic calculation function cannot be called a theory. Muller thinks that it does not mention physical states, physical relations, or physical attributes, and talks little about physical reality outside the laboratory, let alone the microscopic world. To regard it as a "theory" is unacceptable (Muller, 2015: 125).

If a full description of the picture of physical reality is regarded as a necessary condition for a theory, then QM_0, apart from having a certain calculation ability, has not even met the minimum requirements of a physical theory. Muller believes that to interpret QM_0 is to supplement other assumptions and axioms, enrich and modify its theoretical vocabulary, and turn it into a physical theory. The supplementary (or modified) assumptions can be new physical or mathematical structures, or they can just be ontological or epistemological assumptions, such as the attempts of Everett, Bohr, et al.

The third view is put forward by the American philosopher Wallace through a similar analysis (Wallace, 2008; Wallace, 2019). He himself is a famous supporter of the many-worlds interpretation and believes that when teaching standard quantum mechanics in textbooks, the link between eigenstates and eigenvalues and the projection hypothesis should not be taken for granted as "axioms", because these two assumptions are not the fundamental assumptions of quantum mechanics and should not be necessary in the application practice of the quantum mechanics formalism. In fact, their status is highly controversial in different interpretations. Quantum mechanics without these two assumptions is basically the same as the core part defined by Ruetsche and Muller mentioned earlier, and Wallace also believes that this orthodox quantum mechanics is not sufficient to be called a theory, for it provides an effective algorithm that can predict the macroscopic phenomena of recording the results of microscopic measurements, but it is not a theory that can explain the validity of this algorithm (Wallace, 2008: 16–17). Here, the so-called macroscopic phenomena refer to the pointer state of the experimental instrument used for quantum measurement.

In addition, he also divides the interpretations of quantum mechanics into two sorts: one is the so-called pure interpretation, which does not add any new physical or mathematical formalism to the original model, and the other is the "modificatory interpretation", such as the spontaneous collapse dynamics theory, which actually modifies the original physical and mathematical structure of quantum mechanics.

Acuña thus concludes that the "minimal quantum mechanics/core quantum mechanics" mentioned by the above three philosophers are all not complete theories. The interpretation of minimal quantum mechanics/core quantum mechanics is to reform them into complete theories. The various interpretations people talk about are empirically equivalent and competing different theories.

In fact, some researchers in the field of quantum histories interpretation have also discussed this. For example, Hartle, when discussing the relationship between different interpretations of quantum mechanics, pointed out that a formulation that cannot reproduce the standard textbook solution to the probability of quantum measurement is not an interpretation of quantum mechanics. On the contrary, it is a different theory (Hartle, 2005: 73). Kent, the British physicist who has long been concerned with and criticized the quantum histories interpretation, once mentioned that when physicists consider the question of "what is the meaning of giving a new interpretation to quantum mechanics", there are probably two opinions in their minds. One is to understand what quantum mechanics means in an abstract idealistic sense; the other idea is that the ultimate goal should be to pursue scientific progress in a more concrete sense, which includes obtaining new detectable theories, accommodating new calculations, making new predictions, and also inheriting the successful parts of the Copenhagen interpretation – that is, to surpass the Copenhagen interpretation just as general relativity surpasses Newtonian gravity theory. However, the reason why there is such serious controversy and confusion in quantum mechanics is precisely because this topic has become a potential battlefield where two research motivations conflict, and even the authors themselves do not always clearly distinguish between them. (Kent, 1998: 81) It can be seen that the difference between the two that Kent mentioned is precisely the distinction between interpreting an existing theory and constructing a new theory from an existing one.

Losada and Lombardi also made a distinction between theory and interpretation (Losada and Lombardi, 2018: 379). They believe that a theory is composed of a formal structure and its semantic interpretation, the latter giving physical content to the formal structure. The interpretation of semantics also has two levels: the part of "minimal interpretation" applies a formalism to describe the most fundamental physical empirical phenomena in the world, and the other level is the "interpretation" of the theory that is usually referred to, which provides the theoretical elements that are not accessible by experience/experiment. Just as the common practice in the literature of quantum mechanics, the formalism of this theory and its minimal interpretation can be regarded as the core part of the theory, and the interpretation is used to represent the rest. Losada believes that the quantum histories theory can be regarded as a complete theory, including its theoretical core and the interpretive content it is equipped with to solve the traditional problems in the interpretation of quantum mechanics.

Beyond the philosophical arguments about theories and interpretations in science, from the perspective of the history of the development of science, especially the dynamic relationship between the representation in scientific theories and the physical reality, it can also be understood why the quantum mechanics revolution is still an unfinished revolution and why quantum mechanics still needs interpretation. Since the 17th-century scientific revolution, the conflicts and tensions among mathematical representation, experimental observation, and the reality presupposed by theories have constituted the internal driving force for the development and transformation of scientific theories – among which the tension between mathematical representation and physical reality is one of the fundamental driving forces of the physics revolution (Cohen, 2010; Hao, 2014, 2021a,2021b, 2024). The reason why the quantum mechanics revolution is still unfinished is that though people can now conduct high-precision quantum experiments and have a complete system for representing states and operators, there is still no consensus on the ontology presupposed by quantum mechanics and on the reality of quantum states.[2] For relativistic quantum physics and infinite-dimensional quantum theory (such as quantum field theory), it is even difficult to find complete mathematical tools to establish a strict mathematical representation for it, and its ontological commitments are highly controversial (see Kuhlmann, 2023; Kronz and Tracy, 2025). The above discussion views quantum mechanics from the perspective of the dynamics of the development of scientific theories, but it also confirms the previous view on the interpretation of physical theories from the perspective of the philosophy of science.

In summary, this book holds that the meanings of the three terms "theory", "interpretation", and "formalism/formulation" should be more clearly defined. In physics literature, what philosophers call "interpretations" are often directly referred to as "theories" or "formalisms", while in philosophical literature and some physics literature, what is generally called "interpretation" actually refers to "theory" in some cases. For example, terms like "Copenhagen interpretation", "Copenhagen formulation", "many-worlds theory", "many-worlds formulation", "many-worlds formalism", "relative state interpretation", "Bohm theory", "Bohm mechanics", and "hidden variable interpretation" are often used interchangeably and confusingly. In the literature of quantum histories interpretation, there are also several coexisting terms, such as "quantum histories interpretation", "quantum histories theory", "quantum histories formulation", and "quantum histories formalism", with particular confusion.

Therefore, it is necessary to extract clearer definitions and provide a universal reference frame, which will help form a consensus on basic terms and clarify the research ideas and goals. Based on the discussions of Acuña, Ruetsche, Muller, Kent, Losada, et al., the following definitions are attempted for these three terms:

A complete "physical theory" is a mathematical formalism that has received a physical interpretation, mainly composed of a formalism and interpretive content. The formalism can be roughly divided into two parts: first, the quantitative mathematical tools of the theory, including the mathematical structures, mathematical objects, and specific algorithms used, and if non-classical logic is employed, the new logical rules and the corresponding mathematical adjustments need to be clarified; second, it indicates the correspondence between these mathematical structures and objects and the results of observable physical quantities and the physical states of the system and points out to what extent they represent the empirical data involved in the theory.

Interpretive content is the physical interpretation of the above formalism, including clarifying the ontological commitments of the theory, clarifying the connections between the ontologies (laws of nature) and the second-order and higher-order properties of these connections (for example, whether they are deterministic).[3] In addition, it can also include the review of consistency, completeness, and other aspects of formalism. Further speaking, a sufficiently perfect interpretation needs to reveal a world picture with clarity, consistency, completeness, and definiteness: clarity means that the formalism of the theory should satisfy mathematical strictness as much as possible, and the applicable scope and boundary of the theory should be as clear as possible; consistency means that there should be no logical contradiction within the theory itself and it should also be as consistent as possible with related theories; completeness means that it should be able to give a complete description of the physical phenomena within the applicable scope of the theory; definiteness means the uniqueness and unambiguity of the world picture.[4]

When comparing with the existing theories or interpretation schemes in physics, not all of them can meet the above standards. During the interpretation process, people may find that the original theory is logically inconsistent or empirically incomplete in some aspects, which requires modification and supplementation of the formalism, or even the formation of a new formalism, to make it obtain a good interpretation. Therefore, it can be said that the interpretation of physics (that is, giving interpretive content to the mathematical formalism) is definitely a dynamic process.

From such a definition, the "minimal quantum mechanics/core quantum mechanics" mentioned by several scholars is roughly only the core part of the quantum mechanics formalism. Under this standard, the so-called "Bohm mechanics", "spontaneous collapse theory", and "many-worlds theory" are still not complete "theories". Take the "many-worlds theory" as an example. It is not easy for us to determine what its ontological commitments are, and there are many unclear issues in its description of multiple worlds or world branches. To explain why the observer experiences the universe state branch in a certain preferred basis rather than in other bases, and how the symmetry of this basis is broken, may exceed the axiomatization of standard quantum mechanics (Hemmo and Shenker, 2020: 61–70). Bohm mechanics and spontaneous collapse theory exceed the physical presuppositions of standard quantum mechanics. Either they explain the spontaneous collapse of the wave packet by adding a nonlinear dynamic evolution term to the basic equation of quantum mechanics, or they add nonlocal hidden variables and take the hidden variable equation as a new basic equation. However, as mentioned earlier, the experimental evidence to distinguish them from standard quantum mechanics is still insufficient, so they are all in the stage of waiting for verification, and their formalisms are also incomplete.[5] If the predictions that exceed standard quantum mechanics can eventually be verified by decisive experiments, the corresponding formalisms will also show some aspects that exceed the existing standard structure in mathematical representation.[6] For incomplete formalisms, it is possible and necessary to carry out "local interpretation", which is a manifestation of the mutual complementarity of the exploration of formalism and interpretive content. Any fine-tuning of the formalism will "affect the whole", leading to the need for a corresponding re-presentation of the interpretive content; the reflection of the interpretive content will also "reveal the whole from a part", exposing some unnoticed loopholes in the formalism from the still unclear interpretive details.

Such a definition can also open the space for further expansion within various theoretical schemes and the possibility of mutual connection and complementarity among them. The preliminary and local interpretations carried out by people on incomplete formalism often promote the discussion of the two. For example, for Bohm mechanics and spontaneous collapse theory (even those that can accommodate the many-worlds interpretation), some people attempt to use "primitive ontology" as the starting point to give a certain unified "interpretation". This idea has gradually been accepted by many scholars in the fields of Bohm mechanics and spontaneous collapse theory (Chapter 5 will discuss this topic). Another example is Rovelli's relational quantum mechanics, which attempts to uniformly understand quantum mechanics and general relativity under the condition of not changing the formalism and mathematical structure with a set of relationalist philosophy, taking the relation between physical events and systems as the fundamental presupposition (Rovelli, 1996), and providing a weak realist but anti-monist interpretation, and it is possible to form some kind of integration with the quantum Bayesian interpretation, pragmatist interpretation, information theory interpretation, etc.

Twenty years ago, Stoeger had already presented some criteria of adequacy for the interpretation of quantum mechanics, dividing them into four levels (Stoeger, 2001: 82–84). Briefly speaking, the first level he mentioned is the interpretation at the level of the physical theory itself, clarifying the relationship between physical reality and experimental measurement. The interpretation at this level needs to be based on those experiments that have confirmed the basic aspects of the theory, so there is still considerable interpretative freedom. The second level of interpretation focuses on the internal consistency and coherence of the theory; the third level will consider the scope and boundaries of the theory, in other words, the areas where it is no longer applicable and needs to be expanded and supplemented (such as measurement theory), as well as its connection with other physical theories (external consistency, such as classical mechanics or special relativity). And these indeed significantly restrict the interpretation of quantum theory. The fourth level involves more philosophical, i.e., the epistemological and ontological, aspects. For instance, what kind of knowledge about reality does quantum theory provide, and what is still unknown behind this knowledge (in Stoeger's words, "what goes on 'behind the veil', at the level of the underlying quantum entities themselves")? It is not difficult to see that the fourth level of interpretation Stoeger proposed is more like responding to those fundamental philosophical concerns through the understanding of quantum mechanics. Besides, Stoeger also mentioned two meta-philosophical standards, namely "making the least assumptions", "adding as few entities as possible that are not needed when explaining what we have observed", and adhering to a certain degree of realism, because, after all, we cannot arbitrarily manipulate the quantum reality at our will.

Stoeger's classification is actually compatible with the distinction between formalism and interpretive content proposed earlier in this chapter. As we have mentioned, the interpretation of a theory should exhibit clarity, with its applicable scope and boundary of the theory being as clear as possible. This itself implies simplicity, as overly complicated ontological commitments and concept networks inevitably lead to ambiguity and internal inconsistency. In addition, from the epistemological perspective, interpreting a physical theory means decoding a specific world picture from the relatively abstract mathematical structure and the correspondence between mathematics and observational data. Although this picture may not be equivalent to the potential reality itself, this decoding activity is undoubtedly a cognitive one that will be constantly deepened through the development of interpretation.

It is not difficult to find in the literature related to the quantum histories interpretation that the literature which mainly uses it to resolve various quantum paradoxes generally refers to it as "interpretation", such as the approach of Griffiths, et al.; while in the literature that uses this formalism to analyze physical phenomena, the authors more often call it "theory", such as the approach of Hartle, Gell-Mann, Halliwell, Losada, et al. The following chapters will show that the formalism named "quantum histories interpretation/theory" should actually be regarded as multiple different formalisms due to the different given history family selection conditions, although there will be no contradiction in explaining the same empirical data. Generally speaking, if only considering the formalism containing a specific selection condition, one can consider this formalism to be "complete" to a certain extent, but it still has not been clearly interpreted from the ontological perspective. In most of the relevant literature, except for Griffiths' preliminary discussion, most scholars have ignored this aspect. Chapters 5 and 6 will show that by referring to the recent nomological interpretation of the wave function in quantum mechanics, a historicized ontology of quantum histories theory can be preliminarily outlined.

In summary, this chapter takes the most controversial issue of quantum mechanics interpretation as an example to redefine the relationship between theory and interpretation and their relationship with formalism in the context of physics, through the demarcation of formalism, interpretation, and theory. This is not only to clarify the existing confusion in the literature but also to be beneficial to the methodology of related research. The discussion above serves only as a starting point for future work, such as the analysis of other physical fundamental theories and interpretation issues, the question of whether general relativity needs interpretation, the debate on the interpretation of quantum field theory, and the interpretation issues of thermodynamics. It can also be extended to the philosophical reflection on explanation and interpretation in mathematics and computer science. In a broader sense, it can be used to examine whether there can be interactions and mutual reference between the interpretation of natural science and that of humanity science.

The discussion in this section can also be regarded as a study of the methodology of interpretation. Although this book focuses on the interpretation of physics, the entire analysis shows that even the most typical exact science, physics, faces such difficult and perplexing situations when interpreting. Our understanding of quantum reality and the "book of nature" in the microscopic world is still a long and arduous journey, and perhaps only at the beginning stage.

Chapters 3 and 4 will thoroughly examine the quantum histories formalism with consistency conditions or decoherence conditions as selection conditions. Chapter 6 will discuss more quantum histories selection conditions, so the text there will distinguish between the quantum histories formalism and its "interpretation".

Notes

1 Ruetsche's test was put forward with classic physics as a model, and it implies a certain non-contextual semantic realism behind it. Therefore, in her analysis, the attempt directions like Bohm mechanics, spontaneous collapse theory, and many-worlds theory might pass this test, while those interpretations that retain the contextuality truth view of quantum mechanics would find it hard to pass the test.

2 Attempts to modify standard quantum mechanics need to propose reasonable and feasible decision experiments to distinguish the differences between their theoretical predictions and those of standard quantum mechanics. Perhaps due to technical limitations, no experimentally recognized results have been obtained that can be used to refute the usual quantum mechanics.

3 Curiel proposes the distinction of so-called "concrete interpretation", "categorical interpretation", and "metalinguistic interpretation". Concrete interpretation is equivalent to the formalism as in my term, which meets the minimum requirements of instrumentalism and operationalism; categorical interpretation refers to the discussions about the essence of the theory, such as "whether it is deterministic", which are the questions any physical theory has to face; metalinguistic interpretation is the semantics-preserving mapping that maps the theoretical framework to be interpreted to an easily understandable theoretical framework (Curiel, 2009: 46–49).

4 This uniqueness is also referred to as categoricity, which means that all models of the theory are equivalent in a certain isomorphism. See Toader's discussion on whether quantum mechanics has categoricity (Toader, 2021).

5 Acuña holds that, as Bohmian mechanics does not directly appeal to Hilbert space, the nomology within it must be partially sufficient to determine its physical model. Therefore, it has already been interpreted in the sense of Ruetsche's "original picture" and has become a theory (Acuña, 2019: 1721).

6 On this issue, people have different expectations. For instance, Dieks holds that the several mainstream interpretations of quantum mechanics "are empirically equivalent, and no future experimental measurement that uses the well-established interactions that have been widely accepted can break this point" (Dieks, 2017: 307).

Ruetsche now believes that the diversity of ontologically different but observationally equivalent interpretations is inevitable. The deficiency of all the interpretations regarding the "superempirical" virtues (such as simplicity and consistency) leads to the inability to form a unified interpretation. Of course, this is not a defect, because the success of a theory does not necessarily point to a single true picture, but may exactly stem from this diversity and adaptability of interpretations. She advocates giving up the pursuit of a single interpretation of quantum mechanics and instead choosing locally applicable interpretations based on specific scientific practices, emphasizing the murkiness of theories as a scientific virtue. The presupposition of a unified interpretation of scientific theories in scientific realism should be replaced by a locavore philosophy (Ruetsche, 2024).

3 The histories interpretation of quantum mechanics

This chapter discusses the main points and thought sources of the quantum histories interpretation.[1] Section 3.1 introduces its core concepts, theoretical assumptions, and basic formalism first, as well as how probability calculations and propositional reasoning are performed here, and shows how this interpretation depicts the process by which (quasi-)classicality is derived from quantum histories. Section 3.2 sorts out the five important thought sources from which it emerges: the Copenhagen interpretation, Feynman's sum-over-histories formulation of quantum mechanics, Wigner's measurement theory, Everett's relative state interpretation or many-worlds interpretation, and the mechanism of decoherence. Their respective correlations, similarities, and differences with the quantum histories interpretation are briefly analyzed. In the appendix to this chapter, three notorious paradoxes showing intractable aspects of quantum mechanics, namely the EPR paradox, Wheeler's delayed choice gedankenexperiment, and Frauchiger–Renner paradox, are formulated via history chain operator analysis in the quantum histories formalism.

3.1 Main points of the quantum histories interpretation

The core idea of the quantum histories interpretation is that we should no longer solve the probabilistic evolution of the system state merely at a single moment, but use "chain operators" to make an overall and sequential description of the quantum system – any possible evolution process can be a history – and then form a history set by combining several sequential descriptions according to certain requirements, supplemented by consistency conditions, decoherence conditions, or other selection conditions. Only by obeying these rules can we ensure that each quantum history (a sequence of events at different times) in the quantum history family that meets the conditions can be legitimately assigned a classical probability. In addition, when conducting reasoning involving quantum histories, certain necessary principles must also be followed, so as to ensure that the reasoning of propositions conforms to classical logic.

The general process of the establishment and development of this interpretation is roughly as follows. When Griffiths proposed the consistent histories interpretation for the first time in 1984, he primarily wanted to seek a logically

DOI: 10.4324/9781003662860-3

consistent answer to the quantum measurement problem and the wave function collapse problem for closed systems, thus drawing on von Neumann's quantum measurement theory, Everett's interpretation, and Wigner's discussions on the quantum measurement problem in the 1960s. He also compared the similarities and differences with several other interpretations at that time. Starting from the late 1980s, Omnès, along the research line proposed by Griffiths, focused on discussing the logical reasoning of probability propositions in quantum histories and referred to his approach as "logical reformulation of quantum mechanics". He also studied how to use the decoherence mechanism to enable quantum histories to explain the problem of the emergence of classicality. Gell-Mann and Hartle's original research interests were mainly in quantum cosmology and particle physics. A bit later than the work of Griffiths and Omnès, they proposed their version of quantum mechanics concerning closed systems (and the universe as a whole) in the late 1980s, namely the decoherent histories interpretation. They borrowed the influential ideas from the decoherence theory and Feynman's sum-over-histories approach, and meantime developed the idea that the evolution of the physical world has multiple possible choices from the many-worlds interpretation. After enduring arguments against critics for many years, both sides of the debate eventually agreed that Griffiths, Hartle, Gell-Mann, and others had reached essentially equivalent interpretations from different starting points. From then on, many scholars from the United Kingdom, the United States, Japan, and South America also joined the research on the quantum histories formalism.

The following will detail the basic formalism of quantum histories and point out several special issues that this interpretation focuses on. In this formalism, events are the primary elements that can construct histories. For instance, "a certain system is in state $|\psi\rangle$ at time t" can be regarded as an event. Events need to be defined by operators. Since the states that a system is in can all be represented by projectors (also as a density matrix $\hat{P} = |\psi\rangle\langle\psi|$), an event is identified with a projector at a certain time t. If a set of events $\{\hat{P}_k\}$ at time t satisfies the following conditions, then this set is said to provide an exhaustive description of the system:

$$\sum_k \hat{P}_k = I$$

For clarity, it is also required that the system can only be in one event at the same time, that is, these projectors must be mutually exclusive (also known as orthogonal):

$$\hat{P}_j \hat{P}_k = \delta_{jk} \hat{P}_j$$

Then, the event set is said to be exclusive. In the literature, only exhaustive and exclusive event sets composed of projectors are usually considered. For any events $\hat{P}_j, \ldots, \hat{P}_k$ in an event set $\{\hat{P}_k\}$ at the same moment, the addition operation

$\hat{P}_j + \ldots + \hat{P}_k$ can be performed, thereby obtaining all possible events under this event set, denoted as $\left[\hat{P}_k\right]$, which contains a total of $2^k - 1$ events. For example, for the event set $\left\{\hat{P}_1, \hat{P}_2, \hat{P}_3\right\}$, we have

$$\left[\hat{P}_k\right] = \left\{\hat{P}_1, \hat{P}_2, \hat{P}_3, \hat{P}_1 + \hat{P}_2, \hat{P}_1 + \hat{P}_3, \hat{P}_2 + \hat{P}_3, \hat{P}_1 + \hat{P}_2 + \hat{P}_3 = I\right\}$$

Therefore, it contains $2^3 - 1 = 7$ events.

Now consider the events of a quantum system evolving through different times: given some times $\left\{t_1, \ldots, t_n\right\}$, the set of them is called the temporal support of a history. Let the Hilbert spaces of the system at these times be denoted as $\mathcal{H}_1, \mathcal{H}_2, \ldots, \mathcal{H}_n$, respectively. Each event can be regarded as a projector $\hat{P}_{\alpha_i}^{t_i}$ in the Hilbert space $\mathcal{H}_i$ corresponding to the time when it occurs. According to the chronological order of these times $t = \left(t_1, \ldots, t_n\right)$, these projectors can be used to construct a history, for example, $\alpha = \left\{\hat{P}_{\alpha_1}^{t_1}, \ldots, \hat{P}_{\alpha_n}^{t_n}\right\}$. The operator obtained by combining these projectors in chronological order through tensor product is

$$\hat{C}_\alpha \equiv \hat{P}_{\alpha_n}^{t_n} \odot \cdots \odot \hat{P}_{\alpha_1}^{t_1}$$

It is called the chain operator (some literature refers to it as "class operator"). Since the tensor product here connects operators temporally, the symbol $\odot$ is adopted for the tensor product for those operators at certain different times, which is distinguished from the common symbol $\otimes$. A history written in this form is called a homogeneous history. Those not admitting this form are called inhomogeneous histories. For example, the histories obtained after performing several Boolean operations on a homogeneous history (such as $\alpha + \alpha'$ or $\alpha \vee \alpha'$) are all inhomogeneous histories.

If the set $\left\{\alpha\right\}$ of some histories formed by given n moments satisfies

$$\sum_\alpha \hat{C}_\alpha = I^{\odot n}$$

where $I^{\odot n}$ represents the time-dependent tensor product of n identity operators I, then the history set $\left\{\alpha\right\}$ is called a (quantum) history family. Two histories in a history family are called disjoint if the projectors at least at one moment are orthogonal, $\hat{P}_{\alpha_k}^{t_k} \perp \hat{P}_{\beta_k}^{t_k}$.

Between different history families defined on the same time support and the same Hilbert space at each moment, the procedure of coarse-graining and fine-graining can also be defined. A history family $\left\{\alpha\right\}$ is called the coarse-graining of another history family $\left\{\beta\right\}$, if at least for one moment t_k, the event $\alpha_{k,\mu}$ that occurs at that moment in $\left\{\alpha\right\}$ is the coarse-graining of the event that occurs at that moment in

$\{\beta\}$, that is, it is obtained by adding together some of the projectors $\beta_{k,v_1},\ldots, \beta_{k,v_i}$ at that moment in $\{\beta\}$

$$\exists t_k \exists \hat{P}^{t_k}_{\alpha_{k,\mu}} \exists \hat{P}^{t_k}_{\beta_{k,v_1}},\cdots,\exists \hat{P}^{t_k}_{\beta_{k,v_i}} \left(\hat{P}^{t_k}_{\alpha_{k,\mu}} = \hat{P}^{t_k}_{\beta_{k,v_1}} + \cdots + \hat{P}^{t_k}_{\beta_{k,v_i}} \right) \quad (i \geq 2)$$

Conversely, the history family $\{\beta\}$ is the fine-graining of the history family $\{\alpha\}$. Among these individual histories that are connected by coarse-graining and fine-graining processes, a partial order relation can also be formed, denoted as $\beta_k \preceq \alpha_k$. Note that two histories are not always directly comparable. For instance, there can be such a case: at time t_j, one history is the coarse-graining of another history, while at time t_k, the former is the fine-graining of the latter.

After the initial state ρ of the system is determined, the probability of occurrence of a history α can be defined. If the history family $\{\alpha\}$ containing it satisfies the following consistency condition:

$$\mathrm{Re}\left(Tr\left(\hat{C}_\alpha \rho \hat{C}^\dagger_{\alpha'} \right)\right) = 0 \qquad \forall \alpha \neq \alpha'$$

Then, such a history family is called consistent. The detailed expression of the trace operation in the Schrödinger picture is

$$Tr\left(\hat{P}^{t_n}_{\alpha_n} U\left(t_{n-1},t_n\right) \cdots U\left(t_2,t_1\right) \hat{P}^{t_1}_{\alpha_1} \rho_0 \hat{P}^{t_1}_{\alpha'_1} U^\dagger\left(t_1,t_2\right) \cdots U\left(t_{n-1},t_n\right) \hat{P}^{t_n}_{\alpha'_n} \right)$$

In the above formula, $U\left(t_k,t_{k-1}\right) = e^{i\frac{H}{\hbar}\left(t_k - t_{k-1}\right)}$ is the time evolution operator, which is determined by the Hamiltonian of the system at the corresponding moment. And the detailed expression of the trace operation in the Heisenberg picture is

$$Tr\left(\hat{P}^{t_n}_{\alpha_n} \cdots \hat{P}^{t_1}_{\alpha_1} \rho_0 \hat{P}^{t_1}_{\alpha'_1} \cdots \hat{P}^{t_n}_{\alpha'_n} \right)$$

If two chain operators can be expressed as follows:

$$\hat{C}_\alpha \equiv \hat{P}_0 \odot \hat{P}_1 \odot \cdots \odot \hat{P}_n$$

$$\hat{C}_{\alpha'} \equiv \hat{P}'_0 \odot \hat{P}'_1 \odot \cdots \odot \hat{P}'_n$$

Then, it is possible to quickly determine whether the two are orthogonal as long as their initial state or final state operators are orthogonal:

$$\hat{P}_0 \hat{P}'_0 = 0 \quad \text{or} \quad \hat{P}_n \hat{P}'_n = 0$$

Then, they are orthogonal: $Tr\left(\hat{C}_\alpha \hat{C}^\dagger_{\alpha'} \right) = 0$. Then, it is not necessary to specifically solve what the two chain operators are, nor to care about what the projector at the intermediate moment is, thus greatly simplifying the judgment.

A single history (represented by its chain operator) is called intrinsically inconsistent if any consistent history family becomes inconsistent once it is included and regarded as an element of the Boolean algebra of the history family. A necessary and sufficient condition for judging that "a certain history $\hat{C}_\alpha$ is intrinsically inconsistent" is

$$Tr\left(\hat{C}_\alpha\left(I-\hat{C}^\dagger_{\alpha'}\right)\right)\neq 0$$

For instance, the history chain operator of a particle is $\hat{C}_\alpha=\left[z^+\right]\odot\left[x^+\right]\odot\left[z^+\right]$, where the time evolution operator U between adjacent moments is I. This is an intrinsic inconsistent history, where z^+ and x^+, respectively, represent the spin of the particle being in the upward direction of the z-axis and the x-axis. Whether it is a history family or a single history, its consistency or inconsistency is closely related to the specific dynamics of the time evolution. If the time evolution operator U' between adjacent moments of the aforementioned $\hat{C}_\alpha$ satisfies

$$U'\left|z^+\right\rangle=x^+$$

$$U'\left|z^-\right\rangle=x^-$$

$$U'\left|x^+\right\rangle=z^-$$

$$U'\left|x^-\right\rangle=-z^+$$

Then, it is completely feasible to construct a consistent history family that contains $\hat{C}_\alpha$. Note that the consistent history family is also referred to as a framework, a set of alternative histories, or a "realm" or a "domain" in different literature. With the preparation of the concepts above, we can define the probability of a history α occurring in a certain framework:

$$P(\alpha)=Tr\left(\hat{C}_\alpha\rho\hat{C}^\dagger_\alpha\right)$$

If the final state ρ' of the system is also known, the expression of the probability of α that simultaneously conforms to the given initial and final states is

$$P(\alpha)=\frac{1}{Tr(\rho'\rho)}Tr\left(\rho'\hat{C}_\alpha\rho\hat{C}^\dagger_\alpha\right)$$

As a fact, the consistency condition is equivalent to "the probability of an event occurring at any time in the chain operator is additive"[2]:

$$p\left(\hat{P}^{t_1}_{\alpha_1},\ldots,\hat{P}^{t_i}_{\alpha_i},\ldots,\hat{P}^{t_n}_{\alpha_n}\right)=\sum_{\mu_i}p\left(\hat{P}^{t_1}_{\alpha_1},\ldots,\hat{P}^{t_i}_{\mu_i},\ldots,\hat{P}^{t_n}_{\alpha_n}\right)$$

Here, $\sum_{\mu_i} \hat{P}^{t_i}_{\mu_i} = \hat{P}^{t_i}_{\alpha_i}$. According to this probability formula, an important difference between homogeneous history and inhomogeneous history can be revealed: for the homogeneous history family $\{\alpha\}$ of orthogonal normalization, whether the consistency condition is satisfied or not, the above probability formula is calculated:

$$\sum_{\alpha} P(\alpha) = 1$$

As for the history family $\{\alpha\}$ composed of inhomogeneous histories, in general cases, there is

$$\sum_{\alpha} P(\alpha) \neq 1$$

The left side may be greater or less than 1, and the equal sign only holds when the consistency condition is satisfied.[3]

Two homogeneous histories within a history family are said to be decoherent, if

$$D(\alpha,\alpha') := Tr\left(\hat{C}_{\alpha}\rho\hat{C}^{\dagger}_{\alpha'}\right) = 0$$

Similar to consistency, a history family is said to be decoherent if

$$D(\alpha,\alpha') = 0, \ \ \forall\alpha \neq \alpha'$$

It is easy to see that the decoherence condition is obviously stronger than the consistency condition.

Two decoherent history families $\{\alpha\}$, $\{\beta\}$ are said to be compatible if their union $\{\alpha\} \cup \{\beta\}$ is also decoherent. An equivalent condition for the compatibility of two history families is that there exists a history family that is a fine-graining of both of them. There are also the following related definitions regarding coarse-graining and fine-graining:

The completely fine-grained projector at a certain moment refers to the complete set of commuting observables in the Hilbert space at that moment.

A history family $\{\alpha\}$ is said to be an extension of another history family $\{\beta\}$ if the time support of $\{\beta\}$ is a subset of the time support of $\{\alpha\}$.

The fully fine-grained set of histories within a certain time interval refers to the set of histories formed by the completely fine-grained projectors at all moments within that interval, where the time support has been maximally extended. Generally speaking, the fully fine-grained set of histories does not satisfy the consistency condition. And the completely coarse-grained history actually equals the identity operator $I^{\odot n}$,erasing all detailed information.

Starting from the binary operation $D(\alpha,\alpha')$ for two histories $\alpha,\alpha' \in \{\alpha\}$ as mentioned above, the decoherence functionals can be further abstractly defined.

A decoherence functional is a binary complex-valued function $D: \alpha \times \alpha \mapsto \mathbb{C}$, satisfying the following requirements:

1 Hermiticity: $\forall \alpha \forall \alpha' D(\alpha, \alpha') = D^*(\alpha', \alpha)$;
2 Positive definiteness: $\forall \alpha \forall \alpha' D(\alpha, \alpha') \geq 0$;
3 Normalization: $\sum_{\alpha} \sum_{\alpha'} D(\alpha, \alpha') = 1$;
4 Cauchy–Schwarz inequality: $\forall \alpha \forall \alpha' \left| D(\alpha, \alpha') \right|^2 \leq D(\alpha, \alpha) D(\alpha', \alpha')$;
5 Superposition principle: $D(\bar{\alpha}', \bar{\alpha}) = \sum_{\alpha' \in \bar{\alpha}'} \sum_{\alpha \in \bar{\alpha}} D(\alpha', \alpha)$, where $\bar{\alpha}'$, $\bar{\alpha}$ are the coarse-grainings of α', α, respectively.

The reason why D is called the decoherence functional is that when the history of the system is coarse-grained, the influence of interference terms will be excluded. This process is similar to the mechanism of decoherence, thereby obtaining classical probability.

In the quantum histories interpretation, one can perform logical reasoning on the history propositions represented by chain operators similar to classical logic. Under certain conditions, this reasoning is completely consistent with classical propositional reasoning and has the following characteristics:

1 From the linearity of the chain operator, one can obtain the Boolean-like properties of quantum history properties:

   Generally, there is

   $$\alpha_1 + \alpha_2 = \alpha_1 \vee \alpha_2 - \alpha_1 \wedge \alpha_2$$

   If the consistency condition is satisfied, $\alpha_1 \wedge \alpha_2 = 0$, so there is

   $$\alpha_1 + \alpha_2 = \alpha_1 \vee \alpha_2$$

   Let $\bar{\alpha}$ represent the negation of α, i.e., "α does not occur", then

   $$\bar{\alpha} = I - \alpha$$

2 The consistency and decoherence of the entire history family have been defined previously. In fact, it can be generalized to directly define consistency and decoherence at the level of individual histories or history propositions. At this time, the following equation can no longer be directly understood as the probability of a history occurring:

   $$W(\alpha) = Tr\left(\hat{C}_\alpha \rho \hat{C}_\alpha^\dagger\right)$$

But we only regard $W(\alpha)$ as some kind of weight of α. A single history is called consistent if

$$W(\alpha)+W(\bar{\alpha})=1$$

This condition is equivalent to

$$\mathrm{Re}\left(Tr\left(\hat{C}_{\alpha}\rho\hat{C}_{\bar{\alpha}}^{\dagger}\right)\right)=0$$

A single history is called decoherent if

$$Tr\left(\hat{C}_{\alpha}\rho\hat{C}_{\bar{\alpha}}^{\dagger}\right)=0$$

Based on these two definitions, the following conclusions can be drawn:

1 If each history in a history family is consistent, then the history family is consistent;
2 If each history in a history family is decoherent, then the history family is not necessarily decoherent;
3 The negation of a proposition representing any consistent/decoherent history corresponds to a consistent/decoherent history;
4 Given two history families $\{\alpha\}, \{\beta\}$, where $\{\alpha\}$ is the coarse-graining of $\{\beta\}$, if α and β describe the same history in their respective history families, then α is consistent/decoherent if and only if β is consistent/decoherent.

3 The definition of conditional probability is

$$P(\alpha_2|\alpha_1)=\frac{P(\alpha_1+\alpha_2)}{P(\alpha_1)}$$

Omnès (1988, 1999) constructed the logical reasoning structure of history propositions:

$$\alpha_1 \Rightarrow \alpha_2 \quad \text{if and only if} \quad P(\alpha_2|\alpha_1)=1$$

$$\alpha_1 \Leftrightarrow \alpha_2 \quad \text{if and only if} \quad \alpha_1 \Rightarrow \alpha_2 \text{ and } \alpha_2 \Rightarrow \alpha_1$$

For two histories within a consistent history family, the following three expressions about conditional probability and proposition implication are equivalent:

$$\alpha_1 \Rightarrow \alpha_2$$

$$P(\alpha_1 \Rightarrow \alpha_2)=1$$

$$P(\alpha_2|\alpha_1)=1$$

With the definitions of logical relations such as "or", "and", and "implies" among history propositions, the following conclusions can be verified to hold true:

If $\alpha_1 \Rightarrow \alpha_2$ and $\alpha_2 \Rightarrow \alpha_3$, then $\alpha_1 \Rightarrow \alpha_3$

$\alpha_1 \Leftrightarrow \alpha_1$

If $\alpha_1 \Rightarrow \alpha_2$ and $\alpha_1 \Rightarrow \alpha_3$, then $\alpha_1 \Rightarrow (\alpha_2 \wedge \alpha_3)$

$\alpha_1 \Rightarrow (\alpha_1 \vee \alpha_2)$

$(\alpha_1 \wedge \alpha_2) \Rightarrow \alpha_1$

If $\alpha_1 \Rightarrow \alpha_3$ or $\alpha_2 \Rightarrow \alpha_3$, then $(\alpha_1 \vee \alpha_2) \Rightarrow \alpha_3$

If $\alpha_1 \Rightarrow \alpha_2$, then $\bar{\alpha}_2 \Rightarrow \bar{\alpha}_1$

It should be emphasized that the reason why the above theorems (similar to those in classical propositional logic) hold true is that all these history propositions belong to the same consistent family. For history propositions that cross distinct consistent history families (even if they represent the same history chain operator within two compatible history families), generally speaking, these theorems do not always hold true. People can construct various examples of contradictory reasoning (see Chapter 4). In order to resolve these contradictory reasonings and the various quantum paradoxes they bring about, Griffiths has established four principles for the selection of the histories framework and the general proposition reasoning within the framework (Griffiths, 2003: 273–283, 2014, 2024):

1 The principle of liberty. Under the condition of ensuring compliance with other rules, people can freely choose any framework to construct the description of a physical system. The quantum histories interpretation does not impose any prior constraints or regulations on the selection of frameworks.
2 The principle of equality. All possible descriptions of a certain system are equivalent. No framework is more "real" than others, and none is specially selected by nature. An analogy can be made with the equivalence of various inertial systems in special relativity.
3 The principle of incompatibility. The single framework rule requires that physicists must not place history propositions from incompatible frameworks in the same quantum description. The process of history proposition reasoning and probability calculation must be carried out within a single consistent framework.
4 The principle of utility. Although the selection of frameworks is free and all frameworks are of equal status, when describing a given quantum system, choosing certain frameworks can be more convenient and useful.

Griffiths believes that in quantum histories interpretation, the description of physical systems, physical processes, and physical events, or any proposition inference, ought to strictly follow the above principles, especially the single framework rule, to avoid contradictory reasoning and the generating of quantum paradoxes (Griffiths, 2024).

In contrast, he points out that classical physics implies a presupposition about ontology and epistemology: the principle of unicity. This principle asserts that there is only one state of affairs in the world, and it is the only one that is real. Any true assertion about the world must be consistent with it (although people are actually unable to provide such a description). Griffiths argues that this principle is contrary to the fundamental spirit of quantum mechanics and the spirit of the quantum histories interpretation.[4] People have to abandon it for the quantum world. However, some scholars believe that adopting his four principles instead of the principle of unicity would face various problems in physics, logic, and ontology. Chapter 4 will discuss the controversies.

Another topic that Omnès, Hartle, and Gell-Mann focused on is how the classical world emerges from the quantum world. In particular, quantum histories interpretation emphasizes that both macroscopic and microscopic physical processes can be fully expressed by quantum histories without the intervention of macroscopic observers/apparatuses in the measurement process. Therefore, the measurement process is just a part of a quantum history. From an ontological perspective, such an approach eliminates the classical-quantum dualism and returns to monism. Because of this, if quantum histories interpretation aims to provide a successful and complete quantum theory, it must explain the mechanism of the emergence of classicality (Rocha, Rickles and Boge, 2022: 1188–1193).

First of all, it should be clear that, as Hartle and Gell-Mann stated, there is no perfect "classical world" that perfectly conforms to classical physics. Even the evolution process of macroscopic objects containing a large number of particles may be disturbed by quantum effects. This is the basic ground on which quantum cosmology is established. Moreover, if we consider the evolution of the entire universe, we must take into account the influence of phenomena such as quantum fluctuations that occurred in its early stages. What we can discuss is only a "quasi-classical" world and the quasi-classical physical processes occurring within it.[5] In addition to explaining the emergence of quasi-classicality, quantum histories interpretation must also explain why macroscopic observers can observe the persistence and stability of these quasi-classical phenomena in the everyday world.

To express quasi-classical behaviors in the language of quantum histories, one needs to first examine the most finely grained history family of decoherent/consistent histories to provide the most detailed description possible under the respective conditions. Generally speaking, the "maximum" degree of fine-graining here implies that there will be a large amount of quantum coherence, and then a certain degree of coarse-graining must be carried out – this coarse-graining may stem from the actual needs of observational activities or the limitations of the observer's ability. For instance, when examining the position of a particle, it is often only necessary to require it to be within a certain spatial range with a limited degree of precision, rather than precisely at a certain point in space. When examining the macroscopic motion of an material object, it is generally only necessary to consider the momentum and coordinates of the center of mass, without considering all the motion details of the internal particles.

According to Hartle's expectation, people should be able to select an appropriate quasi-classical domain such that the projection decomposition at each moment corresponds to macroscopic properties. This appropriate projection decomposition is the coarse-graining that meets the following conditions: the projectors correspond to large-dimensional subspaces; these projectors are maximally fine-grained, that is, if further fine-grained, they no longer satisfy the consistency conditions. He is convinced that since we live in a definite world governed by classical deterministic laws, in principle, there should be a unique quasi-classical domain that can be found, which contains the history most similar to the classical world's evolution process, and the probability of this history occurring approximately equals to 1.

To explain how people can observe persistent and stable quasi-classical phenomena in the macroscopic world, Gell-Mann and Hartle (1990) introduced an important concept – the information gathering and utilizing system (IGUS for short). Roughly speaking, they are complex adaptive systems that gradually evolve to be able to utilize the relative predictability in the quasi-classical world to acquire information, organize information, and make inferences. Humans, as the observing subjects, are also IGUSes and have evolved to such a degree that they can find a series of compatible frameworks to describe the world, which are sufficient to show the regularities consistent with classical physical laws. This framework that shows quasi-classicality is believed to be not only related to the system's Hamiltonian but also to the initial state of the entire universe. Hartle and others hope to use this concept to help explain why people can perceive a persistent and stable quasi-classical world.[6]

However, the problem lies in that in the quantum histories description, any event or property corresponds to a projector in the history under examination. The quasi-classical domain is no exception and should also correspond to projectors with certain characteristics. The persistence of the quasi-classical domain and its evolution over time should be reflected in these projectors. But this persistence of quasi-classicality is unstable; the quasi-classicality that a system exhibits at some moments in a history does not guarantee its persistence at any subsequent moment, that is, it lacks extensibility to future moments (Dowker and Kent, 1996).

At the current stage, research in this area still remains at the rough outline sketched by Hartle and Gell-Mann. People have not yet found a mathematically precise definition for this concept, nor have they been able to use it to explain complex phenomena involving senses, perceptual structures, and information exchange between IGUSes in detail. Although there have been initial attempts to derive classical physical laws from the quantum histories framework, including discussions on the arrival time problem and hydrodynamic variables (such as Kent, 1996a; Hartle, 2010, 2011; Gell-Mann and Hartle, 2014; Brun and Halliwell, 1996; Halliwell, 2000, 2005; Halliwell and Yearsley, 2009). However, an unavoidable difficulty has yet to be resolved – whether a general method can be found to select those frameworks that can demonstrate quasi-classicality? If such a method does not exist in principle, then all the discussions about IGUS may fall into a circular argument; if it does exist in principle, then the "principle of utility" among Griffith's four reasoning principles can have a more specific and formal expression. More criticisms and doubts will be discussed in Chapter 4, too.

3.2 The sources of ideas of the quantum histories interpretation: a mergence leading to historicized ontology

The basic idea of the quantum histories interpretation did not emerge out of thin air. Through a review of the literature, we can identify its multi-faceted borrowing from previous interpretations and physical achievements. Of course, pointing out the various sources of thought for a theory only indicates which elements of the theory might have been inspired by which physicist. This does not mean that the theory is a simple combination of these ideas. The following examines the five main sources of thought for the quantum histories interpretation.

The first and foremost source of thought is definitely the orthodox interpretation of quantum mechanics – the "Copenhagen interpretation". Currently, the academic community generally believes that many literatures and textbooks present a fusion and generalization of the partial views of the founders of quantum mechanics such as Bohr, Born, Heisenberg, Pauli, and von Neumann, ignoring the differences and contradictions among their thoughts. In fact, there is no consistent and unified "Copenhagen interpretation". This term is actually an "invention" around the 1950s (Howard, 2004). However, we can still roughly summarize the main points of the Copenhagen version of quantum mechanics (some terminologies have multiple meanings, and only the widely used meanings are listed here):

Schrödinger's theory of describing the evolution of the state of a quantum system using the wave function and the wave equation;
Born's probabilistic interpretation of the wave function;
Heisenberg's uncertainty principle (also known as the indeterminacy principle);
Bohr's correspondence principle: the quantum behavior of large-scale and macroscopic systems approaches classical behavior;
Bohr's complementarity principle (complementarity between particle behavior and wave behavior, complementarity between kinematic and dynamic descriptions);
von Neumann's projection hypothesis and its wave packet collapse interpretation;
Bohr, Heisenberg, and others' emphasis on the role of the classical world and classical observers in the quantum measurement process.

It is generally believed that the last five points belong to interpretive content. The formalism of the quantum histories interpretation is clearly different from the Copenhagen interpretation, but as Griffiths and Omnès have stated (Griffiths, 2024; Omnès, 1994: 498), it can be understood as an improvement of the latter (Copenhagen done right), or a "modernized version of the interpretation first proposed by Bohr in his early years". On the one hand, it does not aim to become a new theory to replace quantum mechanics, but rather to provide a consistent and clear expression of the basic content of quantum mechanics; on the other hand, the quantum histories interpretation expresses the ambiguous terms in the Copenhagen interpretation in a mathematical and physical way, eliminates the central role of measurement, and reduces arbitrary and obscure expressions.

How are the core of the Copenhagen interpretation reflected in and updated by the quantum histories interpretation? First, the complementarity principle in the quantum histories interpretation is manifested in the single framework rule, which states that combining histories from incompatible frameworks will bring an inconsistent outcome, that is, for two families of non-commuting projectors, if they are simply merged into one history family for discussion, they will definitely violate the consistency condition and thus we cannot meaningfully calculate the probability of any history therein. In addition, for the calculation of the results of multiple consecutive measurements, the two interpretations are consistent. Starting from the Copenhagen interpretation, the probability formula for a quantum system to successively produce results $\{A_1,\ldots,A_n\}$ after multiple measurements at $\{t_1,\ldots,t_n\}$ can be obtained:

$$P(\alpha_1,\ldots,\alpha_n)=\frac{Tr(A_1\ldots A_n D A_n\ldots A_1)}{Tr(D)}$$

This is consistent with the solution of the probability of consistent histories given earlier. Although the Copenhagen interpretation only regards measurement and measurement results as physically meaningful content, in fact, it selects a small part of all the consistent history families about $S+O$ (the part that is very important and can describe macroscopic experimental results), so of course the two will give the same probability calculation formula here.

However, the Copenhagen interpretation and the quantum histories interpretation also have significant differences. For instance, in the Copenhagen interpretation, quantum mechanics describes any system must have an external macroscopic observer to "measure" it. Then why can't we combine the measured system S and the observer O and regard $S+O$ as a closed system for description? According to the Copenhagen interpretation, measuring $S+O$ must have an external observer O' to observe it, and this can be infinitely inferred. This is the von Neumann chain. Some scholars seek help from the role of the observer's consciousness or try to find a certain "absolute" classical instrument, but these explanations seem to transfer the existing difficulties to new entities that are difficult to clearly interpret. Moreover, the measurement process itself is not the result of the time evolution of the Schrödinger equation, and obtaining the specific measurement result of a non-orthogonal state requires introducing some additional physical mechanism – wave packet collapse. The existence of wave packet collapse means that the Schrödinger equation is insufficient to completely describe quantum processes and also implies the time asymmetry in quantum processes. There are also people like von Neumann who believe that although the physical processes occurring during measurement can be characterized by wave functions, it can only give the state of the system after each ideal measurement, and cannot explain the state before measurement or the correlation between multiple physical events occurring at different times in a closed system. These difficulties do not exist in the quantum histories interpretation. Hartle pointed out that the Copenhagen-style quantum mechanics can be understood as an approximation of quantum mechanics about closed systems, and the measurements carried out in it are ideal measurements that "disturb the subsystem as little as possible", so that the probabilities obtained by using the projectors of the entire closed system and the measured subsystem respectively are sufficiently close (Hartle, 2004: 76).

The second source of thought is the sum-over-histories/path integrals formulation created by Feynman, inspired by Dirac's Lagrangian formulation of quantum mechanics (Dirac, 1933; Feynman, 1948). From the viewpoint of quantum histories formalism, it can be seen that Feynman's sum-over-histories formulation is closely related to the sequence of projectors of histories. Feynman's formulation is roughly as follows:

Consider the histories/paths between two points $a = (x_a, t_a)$ and $b = (x_b, t_b)$ in the configuration space. Divide equally the time interval $T = t_b - t_a$ into N segments:

$$\varepsilon = \frac{T}{N}$$

For each moment t_i within, select a point (x_i, t_i), link all the points, and then a "history" is thus obtained. Taking the limit $N \to \infty$, an infinite-dimensional integral is obtained. It can be considered that all these histories occur with equal probability, but the actual differences in their contributions to the propagator $K(a, b)$ depend on their respective phase factors $S[x(t)]/\hbar$:

$$K(x_a, t_a; x_b, t_b) = N \int_{x(t_a)=x_a}^{x(t_b)=x_b} Dx e^{\frac{i}{\hbar}\int_{t_a}^{t_b} \hat{H}(x)dt} = \sum_{\text{All paths from } x_a \text{ to } x_b} c \cdot e^{\frac{i}{\hbar}S[x(t)]}$$

This makes the propagator contain all the dynamic information. This expression embodies three aspects: first, the quantization process is given by infinite calculus; second, the quantum fluctuation is reflected by the phase factor; third, the sum over histories embodies the superposition principle of quantum mechanics.

Let's take a simple example to see the connection between Feynman's formulation and the quantum histories formulation. Omnès pointed out that if we consider a history with only two moments, and the two projectors A_1 and A_2 are, respectively, the momentum or position operators of a particle in one-dimensional space, then the description of dynamics by the Feynman sum can directly lead to the expression of probability and the summation rule (Omnès, 1994: 130). Under the Feynman formulation, the probability that the particle is at x_1 at t_0 and at x_2 at t_2 is

$$\langle x_2, t_2 | x_1, t_0 \rangle = \sum_F e^{\frac{i}{\hbar}S_F(x_2, t_2; x_1, t_0)}$$

F denotes all the possible paths from (x_1, t_0) to (x_2, t_2). The probability of finding the particle within a small interval Δx near x_2 at t_2 is $\langle x_2, t_x | x_1, t_0 \rangle^2 \Delta x$. Considering a certain moment t_1 between t_0 and t_2, let $E'(t_1)$ be the projector representing the particle passing through the configuration region $\mathcal{D}'$ at t_1. Then, the matrix elements of this operator (represented in matrix form) are

$$\langle x_2, t_2 | E'(t_1) | x_1, t_0 \rangle = \sum_{F'} e^{\frac{i}{\hbar}S_{F'}(x_2, t_2; x_1, t_0)}$$

F' denotes all the possible paths through the configuration region $\mathcal{D}'$ at t_1. In exactly the same way, the matrix elements of the projector $E''(t_1)$ of the particle passing through the region $\mathcal{D}''$ at t_1 are

$$\langle x_2, t_2 | E''(t_1) | x_1, t \rangle_0 = \sum_{F''} e^{\frac{i}{\hbar}S_{F''}(x_2, t_2; x_1, t_0)}$$

Consider a sequence of events in which a particle passes through the region $\mathcal{D}'$ at t_1 and then appears in the small interval Δx near x_2 at t_2. The projector representing "occurring in an interval Δx near x_2" is

$$E(t_2) = \int_{\Delta t_2} |x_2, t_2\rangle\langle x_2, t_2|\, dx$$

Then the probability of this history can be expressed as

$$P = Tr\Big(E(t_2)E'(t_1)|x_1, t_0\rangle\langle x_1, t_0|\, E'(t_1)\Big)$$

This is exactly the probability expression of this history in the quantum histories interpretation. If the vector $|x_1, t_0$ at t_0 is expressed by the normalized wave function ψ written as $|\psi, t_0$, then the above probability expression is

$$P = Tr\Big(E(t_2)E'(t_1)|\psi, t_0\rangle\langle\psi, t_0|\, E'(t_1)\Big)$$

We could say that quantum histories are actually a kind of Feynman history. If generalized to more time and higher dimensional space, the analysis process remains similar.[7] The above expression shows good additivity: if the event of t_1 in the above history is changed to be "the particle passes through the region $\mathcal{D}'$ at t_1, or the region particle passes through the region $\mathcal{D}''$ at t_1", then the above two probabilities can be directly added to get the probability:

$$P = \left|\sum_{F'} \exp\frac{iS_{F'}(x_2, t_2; x_1, t_0)}{\hbar} + \sum_{F''} \exp\frac{iS_{F''}(x_2, t_2; x_1, t_0)}{\hbar}\right|^2$$

$$= \left|\sum_{F'} \exp\frac{iS_{F'}(x_2, t_2; x_1, t_0)}{\hbar}\right|^2 + \left|\sum_{F''} \exp\frac{iS_{F''}(x_2, t_2; x_1, t_0)}{\hbar}\right|^2$$

So, if the probability expression is required to satisfy additivity, then the histories that are selected by this requirement are those where both the probabilities and amplitudes can be added. Furthermore, the decoherent functionals between t_0 and t_f can be rewritten as (Zhang, Dahlsten and Vedral, 2020: 9):

$$D([E],[E']) = \int_{[E]} \mathcal{D}(q_i) \int_{[E']} \mathcal{D}(q_i')\exp\big(iS[q_i] - iS[q_i']\big)\delta\big(q_{i,f}, q_{i,f}'\big)\rho\big(q_{i,0}, q_{i,0}'\big)$$

This simple example in non-relativistic quantum mechanics clearly demonstrates the connection between quantum histories and Feynman histories.[8]

The third source of ideas is the quantum measurement theory proposed by Wigner. *Quantum Theory and Measurement* edited by Wheeler J. A. and Zurek W. H. contains three papers by Wigner discussing quantum measurement and the mind–matter relation. Among them, *The Problem of Measurement* evaluates the treatment of the measurement problem by the Copenhagen interpretation and its difficulties and discusses the meaning of the state vector and the probability solution of multi-time quantum measurement (Wigner, 1963), which has a direct impact on Griffiths' original work.[9]

When discussing the true meaning of the state vector, Wigner believes that, according to quantum mechanics, all obtainable information is obtained through measurement results. The standard way to obtain the state vector is to perform measurements on the system. If a small system interacts with a large system in a definite state and the state of the large system is known, then the small system will also be in a nearly definite state. Therefore, if a hydrogen atom in an excited state is placed in a huge container without radiation, it will quickly transfer all its energy to the radiation field of the container and return to the ground state.

To answer the question of what the state vector is, first, examine the probability calculation formula for successive measurements on a system and obtain definite results (the formulations in both the Schrödinger picture and the Heisenberg picture are given below). Suppose that n successive measurements were made on a certain system at times $t_1, t_2, \ldots, t_n$. In the Schrödinger picture, the operators corresponding to these n measurement processes are denoted as $Q_1, Q_2, \ldots, Q_n$, and the eigenvectors and eigenvalues corresponding to each measurement are represented by ψ and q (all subscripts below represent indices), for example

$$Q_j \psi_k^j = q_k^j \psi_k^j$$

In the Heisenberg picture, the state remains unchanged while the operators vary with time. The operator at time t_j can be expressed in terms of the operator at the initial time:

$$Q_j^H = e^{iHt_j} Q_j e^{-iHt_j}$$

At this point, the eigenvector is written as

$$\phi_k^j = e^{iHt_j} \psi_k^j$$
$$Q_j^H \phi_k^j = q_k^j \phi_k^j$$

If the initial state vector is $\boldsymbol{\Phi}$, then the probability of obtaining a series of results $q_\alpha^1, q_\beta^2, \ldots, q_\nu^n$ after n measurements is the square of the absolute value of the following expression:

$$\left(e^{-iHt_1}\boldsymbol{\Phi}, \psi_\alpha^1\right)\left(e^{-iH(t_2-t_1)}\psi_\alpha^1, \psi_\beta^2\right)\cdots\left(e^{-iH(t_n-t_{n-1})}\psi_\mu^{n-1}, \psi_\nu^n\right) \tag{3.1}$$

This equation in the Heisenberg picture is more concise:

$$\left(\boldsymbol{\Phi}, \phi_\alpha^1\right)\left(\phi_\alpha^1, \phi_\beta^2\right)\cdots\left(\phi_\mu^{n-1}, \phi_\nu^n\right) \tag{3.2}$$

Here, the probability is not solely determined by the n operators $\{Q_j^H\}$ and their eigenvectors in the Heisenberg picture. The temporal sequence of the measurement process also has an essential impact on the outcome. Von Neumann has derived these expressions and extended them to the case where each eigenvalue corresponds to multiple eigenvectors (i.e., degeneracy exists) (von Neumann, 1932 [2018]). In the presence of degeneracy, for each eigenvalue q_j of the Heisenberg operator Q_j^H, a projector P_{jk} is introduced, and thus the probability that the measurement results are in sequence $q_\alpha^1, q_\beta^2, \ldots, q_\nu^n$ is

$$\left(P_{n\nu} \cdots P_{2\beta} P_{1\alpha} \Phi, P_{n\nu} \cdots P_{2\beta} P_{1\alpha}\right)$$

To obtain equations (3.1) and (3.2), one can also assume that when the measurement of Q^j yields the result q_k^j, the state vector becomes ϕ_k^j. In fact, the statement "the state vector is ϕ_k^j" merely means that after the last measurement of the system Q^j, we obtained the result q_k^j. When the eigenvalues are non-degenerate, the final state of the state vector only depends on the last measurement, and the future behavior of the system is independent of the history state at the nearest moment. However, if q_k^j is degenerate, the situation may not be so.

After normalization, the "simplest expression" of the final state vector in the Heisenberg picture is as follows:

$$P_{jk} \cdots P_{2\beta} P_{1\alpha} \Phi \tag{3.3}$$

If after normalization, equation (3.3) is independent of the initial state vector Φ, it indicates that the number of measurements is sufficient to completely determine the state of the system and produce a pure state. If the vector in equation (3.3) still depends on Φ and no specific initial state is given, the system is in a mixed state of all possible Φ under equation (3.3).

Based on the above discussion, Wigner believed that the state vector is merely an abbreviation that contains part of the information we know about the system's past, and this information is related to the prediction of the system's future behavior. The density matrix also has a similar role, but the density matrix may not be able to completely "determine" the future behavior of the system as the state vector does (for example, when the density matrix represents a mixed state). The laws of quantum mechanics yield the probability connection between the results of successive observations on a system, and this connection is only derived in the sense of probability.

The above are the key points of Wigner's discussion on quantum measurement. The influences of these views upon the quantum histories interpretation are as follows: (1) Wigner emphasizes the dependence of the probability formula for multiple measurements on the measurement sequence, and the sequence plays an essential role in the expression; (2) The probability expression given by Wigner here is precisely the probability expression of a certain history occurring after the quantum history family satisfies the consistency condition.

The fourth source of thought is Everett's relative state interpretation and the subsequent development of the many-worlds interpretation. The relative state interpretation does not change the formalism of orthodox quantum mechanics, removes the collapse assumption and the classical observer outside the system, achieving a "physical simplicity" in the ontological sense, but the cost is the commitment to the existence of relative states or multiple worlds (DeWitt and Graham, 1973).

The Everett interpretation holds that only unitary quantum mechanics dynamics is needed without a collapse mechanism to completely describe quantum physics; but at the same time, a significant change in the understanding of the concept of quantum state is required. Assuming that the universe is in a pure state, then any subsystem also has its own quantum state; given the initial state of the universe at time t, the state of each subsystem of the universe with a non-zero amplitude is determined as a "relative state", relative to the rest of the universe at t.

The basic statement of Everett's theory is as follows: Suppose the Hilbert space describing the quantum state of the universe is $\mathcal{H}$, and the universe is initially in a pure state $|\Psi\rangle$. If the universe is divided into two subsystems, $\mathcal{H} = \mathcal{H}_1 \otimes \mathcal{H}_2$, representing the subsystem to be examined and "the rest of the universe", with bases $\{|\psi_1\rangle\}$ and $\{|\phi_2\rangle\}$, respectively, then the quantum state of the universe can be rewritten as

$$|\Psi\rangle = \sum_{ij} \lambda_{ij} |\psi_i\rangle \otimes |\phi_j\rangle$$

where $\sum_{ij} \lambda_{ij}^2 = 1$. However, since $|\Psi\rangle$ generally cannot be directly decomposed into the tensor product of pure states $|\psi_i\rangle \otimes |\phi_j\rangle$, both subsystems $\mathcal{H}_1$ and $\mathcal{H}_2$ are generally in mixed relative states. Define

$$|E_i\rangle = \sum_j \frac{\lambda_{ij}}{\sqrt{\mu_i}} |\phi_j\rangle$$

where

$$\mu_i = \sum_j |\lambda_{ij}|^2$$

Then $|\Psi\rangle$ can be expressed as

$$|\Psi\rangle = \sum_i \sqrt{\mu_i} |\psi_i\rangle \otimes |E_i\rangle$$

So, the relative state of the mixed system $\mathcal{H}_2$ is $\rho_2 = \sum_i \mu_i |E_i\rangle\langle E_i|$, where each $|E_i\rangle$ represents the state of the rest of the universe relative to the system state $|\psi_i\rangle$. Everett referred to each term $|\psi_i\rangle \otimes |E_i\rangle$ as a branch (DeWitt and some others call it a "world"). If one considers actual physical observations and decomposes $|\Psi\rangle$ into parts such as the system to be observed $|\psi_i\rangle$, the instrument $|D_i\rangle$, the observer $|O_i\rangle$, and the environment $|E_i\rangle$, it can be written as

$$|\Psi\rangle = \sum_i c_n |\psi_i\rangle \otimes |D_i\rangle \otimes |O_i\rangle \otimes \cdots \otimes |E_i\rangle$$

When Hartle and Gell-Mann established their generalized quantum mechanics/decoherent histories interpretation, they explicitly called their version a "post-Everett" theory for describing closed systems, which was an extension, clarification, and improvement of the Everett interpretation (Gell-Mann and Hartle, 1990). Their theoretical starting point was the same as Everett's – aiming to provide a coherent description of the entire universe, solve the conceptual difficulty that quantum measurement must rely on external observers, and hope to explain how classical objects and phenomena in the macroscopic world are derived from the quantum world, rather than simply postulating them as the premise of the theory as in the Copenhagen interpretation. They pointed out that the greatest deficiency of the Everett interpretation was: first, it failed to explain how the classical world emerged from quantum phenomena; second, it failed to clearly explain the meaning of "branching"; third, it discarded the probability in quantum mechanics, which has led many people to have doubts about whether probability exists in the many-worlds interpretation and, if so, how it is defined. They believed that, in fact, Everett's theory could also be regarded as a many-histories interpretation – each consistent history family (they called it "multiple alternative histories of the universe") in the quantum histories interpretation was equivalent to an Everettian branch/world. And the way Everett and DeWitt et al. expressed it as "equally real" multiple worlds was misleading. The essence of this statement was merely to argue that quantum mechanics does not give any branch/history family a special ontological status.

If the ontological differences are ignored, from the perspective of formalism, the quantum histories interpretation can be regarded as an extension of the "single-time" many-worlds interpretation to a multi-time one. Hartle and Gell-Mann's early related papers also clearly indicated the connection between the two ideas, calling the decoherent histories interpretation a Neo-Everettian interpretation. Gell-Mann claimed that a given system can have different histories, and each universe's history has its probability of occurrence; so there is no need to make people confusedly accept multiple parallel universes of equal reality. W. H. Zurek summarized the quantum histories interpretation as adding three key mechanisms to the many-worlds interpretation: the concept of multiple alternative coarse-grained histories for a quantum system; the decoherence effect within a selected set of histories; and the approximate determinism toward the classical limit (Zurek, 1991: 36).

The quantum histories interpretation does not accept the Everettian determinism but holds that the nature of quantum mechanics is probabilistic. When discussing the usual small-scale physical processes, there is no need to resort to the universal wave function, nor is there a need for the splitting of the universal state and the coexistence of various macroscopic states after the split. This splitting in the quantum histories interpretation can be understood as follows: an initial state can lead to multiple different macroscopic states subsequently, and under the condition of consistency, each situation occurs according to a meaningful probability. This is similar to the situation in classical statistical physics: starting from an unstable initial macroscopic state, the system can evolve to different macroscopic states, but here the meaning of "different macroscopic states" refers to the system finally being in one of these possible states with a certain probability, or that the evolution of a large number of systems in an ensemble will cause them to be distributed in different macroscopic states according to this probability, rather than referring to the splitting of multiple systems from an independent system.

Although the analogy with classical statistical physics mentioned above does not imply that the logic of quantum mechanics is exactly the same as classical logic, within the same consistent history family, the distribution and calculation of probabilities are consistent with the rules in classical physics (Boolean operations). Therefore, it is quite natural to interpret the splitting of each history branch within the same consistent history family by analogy with the understanding of classical statistical physics.

The fifth source of thought is the decoherence theory. The decoherence theory was first proposed by German physicist Dieter Zeh and others in the 1970s and was later vigorously promoted and developed by theoretical physicists and experimental physicists such as Joos and Zurek.

The decoherence effect refers to the fact that a quantum system is inevitably in interaction with the environment. If a certain observable is measured, the interference terms in the superposition state composed of multiple eigenstates of the observable will fade due to the interaction with the environment. A simple description of this process is as follows: Let the lower subscript S represent the system and E represent the environment. Then, the density matrix of the system and the environment after coupling is

$$\hat{\rho}_S = Tr_E\left(\hat{\rho}_{S,E}\right) = \sum_{i,j} \alpha_i \alpha_j^* \left|S_i\right\rangle\left\langle S_j\right| Tr\left(\left|E_i\right\rangle\left\langle E_j\right|\right) = \sum_{i,j} \alpha_i \alpha_j^* \left|S_i\right\rangle\left\langle S_j\right| \left\langle E_j \middle| E_i\right\rangle$$

Due to the fact that the orthogonal terms of the environmental part, $\left\langle E_j \middle| E_i \right\rangle \approx 0$, for $i \neq j$, all the off-diagonal terms of the entire expression rapidly vanish, and it eventually becomes $\hat{\rho}_S \approx \sum_i \left|\alpha_i\right\rangle^2 \left|S_i\right\rangle\left\langle S_i\right|$.

The environment referred to here can not only be the environment outside the quantum system in the usual sense but also an internal environment formed by particles within the system; the quantum system mentioned here is not only a multi-particle system or macroscopic system that can be manipulated in the laboratory but also applicable to the entire universe (if a quantum mechanical description of the entire universe is to be made). It is precisely because decoherence is ubiquitous that we can observe the stable existence of macroscopic physical systems, rather than observing them in a superposition state; of course, this decoherence that occurs at an extremely rapid speed is a serious obstacle for quantum computing and quantum communication that essentially rely on quantum entanglement and quantum coherence. If the state of the system is expressed by the density matrix, then the decoherence effect refers to the fact that the influence of interaction with the environment will cause the off-diagonal terms of the density matrix to basically disappear. The resulting density matrix without off-diagonal terms indicates that the system is finally in a mixed state, with each branch having a classical probability distribution. A simplified description can represent the process of decoherence and collapse as follows:

$$\begin{bmatrix} p_1 & \rho_{12} \\ \rho_{21} & p_2 \end{bmatrix} \overset{\text{decoherence}}{\Rightarrow} \begin{bmatrix} p_1 & 0 \\ 0 & p_2 \end{bmatrix} \overset{\text{collapse}}{\Rightarrow} \begin{cases} \begin{bmatrix} p_1 & 0 \\ 0 & 0 \end{bmatrix} \\ \begin{bmatrix} 0 & 0 \\ 0 & p_2 \end{bmatrix} \end{cases}$$

The decoherence effect itself does not have a direct correspondence with various interpretations of quantum mechanics, and it does not exclusively support one interpretation while excluding others. However, it emphasizes the interaction between the environment and the system as well as the process of coherence transitioning from being present to absent. It also helps in explaining the status of classical objects in the Copenhagen interpretation, explaining the issue of preferred basis vectors and clarifying the concept of "branches" or "worlds" in the many-worlds interpretation.

Some characteristics of the decoherence mechanism have also inspired Hartle and Gell-Mann's proposal of the decoherent histories interpretation. They once mentioned that the research on decoherence by Zeh, Zurek, and others (Zeh, 1970; Zurek, 1982; Joos and Zeh, 1985) and the quantum histories formulation by Griffiths are the basis for the interpretation of decoherent histories. The decoherence mechanism does not go beyond standard quantum mechanics and is not a modification of it, but the study of it has broken some philosophical dogmas set by the Copenhagen interpretation, such as the necessity of the observer's intervention in the disappearance of coherence, and it can naturally explain why macroscopic objects exhibit spatial locality and why quantum coherence does not occur macroscopically. In addition, Hartle and Gell-Mann envisioned giving a specific derivation mechanism for the quasi-classical domain through the interpretation of decoherent histories, and the decoherence mechanism is one source of this vision (see the relevant analysis in Chapter 6).

The above has sorted out the five major sources of thought of the quantum histories interpretation. To summarize, the quantum histories formalism is significantly different from the presentation given in textbooks, but it is not an empty castle in the air. Rather, it takes its shape from key ideas of the well-known interpretations and physical mechanisms at that time. Subsequent chapters will make clear that the quantum histories formalism brings an ontological picture that is different from the usual quantum mechanics concepts. The analysis here shows that, unlike those approaches that take physical modifications into quantum mechanics, this new picture given by the quantum histories interpretation is basically rooted in standard quantum mechanics itself. It means that even in the case of non-modified quantum mechanics, a deeper understanding of the essence of quantum reality is still far from finished.

Appendix: Cases studies: EPR paradox, Wheeler's delayed choice gedankenexperiment, Frauchiger–Renner paradox in histories formalism

In this appendix, we will focus on handling the three most typical "paradoxes" – EPR paradox, Wheeler's delayed choice gedankenexperiment, and Frauchiger–Renner paradox – by using the quantum histories formalism, so as to demonstrate the distinctive approach and advantages of this formalism.

First, consider the EPR paradox. Let there be two particles 1 and 2 with spin-$\frac{1}{2}$, each having two directions denoted as $|\uparrow_1\rangle$, $|\downarrow_1\rangle$, $|\uparrow_2\rangle$, $|\downarrow_2\rangle$. Without loss of generality, let the angle between the two upward spin directions be denoted as θ.

From this, it is not difficult to obtain that the four quantities satisfy the following relation:

$$\begin{cases} |\uparrow_2\rangle = \cos\frac{\theta}{2}|\uparrow_1\rangle + \sin\frac{\theta}{2}|\downarrow_1\rangle \\ |\downarrow_2\rangle = -\sin\frac{\theta}{2}|\uparrow_1\rangle + \cos\frac{\theta}{2}|\downarrow_1\rangle \end{cases} \quad \text{3.A1)}$$

Consider two spin-measuring devices, D_A and D_B, which measure the z-directional spins of particles 1 and 2, respectively. At the initial moment t_0, the state of the entire system, including the measuring devices, can be denoted as

$$|\Psi\rangle_0 = \frac{1}{\sqrt{2}}\left(|\psi_{D_A}\psi_{D_B}\uparrow_1\downarrow_1\rangle + |\psi_{D_A}\psi_{D_B}\downarrow_1\uparrow_1\rangle\right)$$

Of course, based on (3.A1), $|\psi_0\rangle$ can also be expressed in terms of $|\uparrow_2\rangle$ and $|\downarrow_2\rangle$. At some subsequent time t_1, particle 1 is observed by D_A; at a later time t_2, particle 2 is observed by D_B. Consider the following exhaustive and exclusive decomposition $\mathcal{A}_{1,2}$ of the identity in the Hilbert space:

$$\mathcal{A}_1 = \begin{cases} |\psi_{D_A,\uparrow_1}\psi_{D_B}\uparrow_1\uparrow_2\rangle\langle\psi_{D_A,\uparrow_1}\psi_{D_B}\uparrow_1\uparrow_2| \\ |\psi_{D_A,\uparrow_1}\psi_{D_B}\uparrow_1\downarrow_2\rangle\langle\psi_{D_A,\uparrow_1}\psi_{D_B}\uparrow_1\downarrow_2| \\ |\psi_{D_A,\downarrow_1}\psi_{D_B}\downarrow_1\uparrow_2\rangle\langle\psi_{D_A,\uparrow_1}\psi_{D_B}\downarrow_1\uparrow_2| \\ |\psi_{D_A,\downarrow_1}\psi_{D_B}\downarrow_1\downarrow_2\rangle\langle\psi_{D_A,\uparrow_1}\psi_{D_B}\downarrow_1\downarrow_2| \end{cases}$$

$$\mathcal{A}_2 = \begin{cases} |\psi_{D_A,\uparrow_1}\psi_{D_B,\uparrow_2}\uparrow_1\uparrow_2\rangle\langle\psi_{D_A,\uparrow_1}\psi_{D_B,\uparrow_2}\uparrow_1\uparrow_2| \\ |\psi_{D_A,\uparrow_1}\psi_{D_B,\downarrow_1}\uparrow_1\downarrow_2\rangle\langle\psi_{D_A,\uparrow_1}\psi_{D_B,\downarrow_1}\uparrow_1\downarrow_2| \\ |\psi_{D_A,\downarrow_1}\psi_{D_B,\uparrow_2}\downarrow_1\uparrow_2\rangle\langle\psi_{D_A,\uparrow_1}\psi_{D_B,\uparrow_2}\downarrow_1\uparrow_2| \\ |\psi_{D_A,\downarrow_1}\psi_{D_B,\downarrow_2}\downarrow_1\downarrow_2\rangle\langle\psi_{D_A,\uparrow_1}\psi_{D_B,\downarrow_2}\downarrow_1\downarrow_2| \end{cases}$$

Here, $\psi_{D_A,\uparrow_1}$ represents that D_A observed particle 1 in the state $\uparrow_1$, and the meanings of the other notations follow accordingly. From this, the following family of histories can be obtained:

$$\mathcal{F}: |\psi\rangle_0 \rightarrow [\mathcal{A}_1] \rightarrow [\mathcal{A}_2]$$

In this decoherent family, we can assign and calculate the probability for each of the histories. The remark I would like to make here, regarding the approach of using decoherent histories to handle the EPR paradox, is that the analysis of decoherent histories shows that this kind of correlation between particles does not require physical action at a distance as a prerequisite. We can explain this paradox with a local theory and examine the entire process in a closed system without an observer. The measurement device is treated as part of the entire system. Of course, for systems without measurement devices, the same probabilistic results can also be obtained using decoherent histories (Cheung, 2020).

The second one, Wheeler's delayed choice gedankenexperiment, or the delayed choice paradox, is a typical instance that manifests the quantumness in physical processes of the microscopic world through temporal retrocausality. To put it simply, the paradox is as follows (Griffiths, 2003: 273–283):

A photon initially enters a Mach-Zehnder interferometer, and its path subsequently bifurcates into two parts. There are two mirrors along each of these two paths, causing the paths to converge at a specific point. If a second interferometer is placed at this convergence point, upon passing through this second interferometer, the photon will finally enter detector D_1 with a probability of 100% due to the phase difference between the two paths. If there is no second interferometer at that convergence point, the photon will enter detector D_1 or detector D_2 with a probability of 50% respectively, depending on whether it takes path 1 or path 2. The paradox lies in the fact that if the second interferometer is not at the convergence point and the photon has already traversed the first interferometer , traveling on path 2 (there is a 50% probability of this situation, let's assume so for now), then, just before the photon reaches the convergence point, if the second interferometer is suddenly inserted, it seems that the photon, which was originally intended to follow path 2, suddenly has to be in a superposition of paths 1 and 2, undergo interference, and ultimately be directed toward detector 1. It appears that a later physical event has, retroactively, influenced the previous physical process: the insertion of the second interferometer has affected whether the photon's path was on a specific path or in a superposition state of paths 1 and 2 earlier!

Let the initial state of the system be $\Omega_0 = |0a\rangle|B_0\rangle|D_1\rangle|D_2\rangle$, where $|0a\rangle$ indicates that the photon is at position a at time 0, and $|D_1\rangle|D_2\rangle$ indicates that neither of the two detectors has detected the photon. The event "the second interferometer is placed in or taken out" (i.e., $|B_{\text{in}}\rangle$ or $|B_{\text{out}}\rangle$) is denoted as B, occurring at time 2, that is, $|B_0\rangle \rightarrow (|B_{\text{in}}\rangle + |B_{\text{out}}\rangle)/\sqrt{2}$. $|D_1'\rangle$ and $|D_2'\rangle$, respectively, represent that D_1 and D_2 have detected the photon. The entire process can be characterized by the following three consistent history families:

The first history family contains three histories, with a bifurcation starting from event B. The history chain operator that has gone through times 0, 1, 2, 3, and 4 is

$$\Omega_0 \odot \left[(1c + 1d)/\sqrt{2} \right] \odot \begin{cases} B_{\text{in}} \odot [3f] \odot D_1' \\ B_{\text{out}} \odot \left[(3e + 3f)/\sqrt{2} \right] \odot \begin{cases} D_1' \\ D_2' \end{cases} \end{cases}$$

c, d, e, f are those forked paths. $(1c+1d)/\sqrt{2}$ represents (that the photon is in) the superposition of two paths split after crossing the first interferometer, and $(3e+3f)/\sqrt{2}$, the superposition after crossing the second interferometer.

The second history family encompasses three histories, which start to branch out from the initial state Ω_0 at time 0:

$$\Omega_0 \odot \begin{cases} \left[(1c + 1d)/\sqrt{2} \right] \odot B_{\text{in}} \odot [3f] \odot D_1' \\ [1c] \odot B_{\text{out}} \odot [3f] \odot D_1' \\ [1d] \odot B_{\text{out}} \odot [3e] \odot D_2' \end{cases}$$

The third history family consists of four histories, with bifurcations taking place twice at moments 1 and 2:

$$\Omega_0 \odot \begin{cases} [1c] \odot \begin{cases} B_{\text{in}} \odot \left[(3e + 3f)/\sqrt{2} \right] \odot \left[\left(|D_1'\rangle |D_2\rangle + |D_1\rangle |D_2'\rangle \right)/\sqrt{2} \right] \\ B_{\text{out}} \odot [3f] \odot D_1' \end{cases} \\ [1d] \odot \begin{cases} B_{\text{in}} \odot \left[(-3e + 3f)/\sqrt{2} \right] \odot \left[\left(-|D_1'\rangle |D_2\rangle + |D_1\rangle |D_2'\rangle \right)/\sqrt{2} \right] \\ B_{\text{out}} \odot [3e] \odot D_2' \end{cases} \end{cases}$$

From these three history families, the entire gedankenexperiment can be analyzed from different perspectives. It is not hard to see that there is no physical event that occurs later but affects a previous process.

The third paradox, known as Frauchiger–Renner argument, is another type of paradox, Wigner's friend type, designed to show the single-world view of quantum mechanics is inconsistent. Since it involves distinct perspectives of observation from distinct observers and also a meticulous arrangement of the moments when the event occurs, the application of the quantum histories formalism is actually very convenient in clarifying the crux of this paradox. Here, we show a simple presentation of the analysis from Losada et al. (2019).

The experimental setup includes two participants. Two "friends" observers (F_1, F_2) perform measurements in isolated laboratories (L_1, L_2), and two external "Wigner" observers (W_1, W_2) measure the entire laboratory thereafter. The quantum state evolution can be summarized as follows: F_1 measures the quantum coin state $|\varphi\rangle = \sqrt{\frac{1}{3}}|h\rangle + \sqrt{\frac{2}{3}}|t\rangle$ and prepares a qubit $|\downarrow\rangle$ or a qubit $|\rightarrow\rangle$ according to the result $|h\rangle$ or $|t\rangle$, respectively, where $|\rightarrow\rangle$ represents the state $\frac{|\uparrow\rangle + |\downarrow\rangle}{2}$. F_2 measures the S_z of the qubit, and it is obvious that the overall system state evolves to $|\Psi\rangle = \frac{1}{\sqrt{3}}|H\rangle|\Downarrow\rangle + \sqrt{\frac{2}{3}}\,|T\rangle|\Rightarrow\rangle$, where $|H\rangle$, $|T\rangle$, $|\Downarrow\rangle$, $|\Rightarrow\rangle$ are the state of the laboratories corresponding to $|h\rangle$, $|t\rangle$, $|\downarrow\rangle$, $|\rightarrow\rangle$.

W_1 and W_2, respectively, perform X and Y basis measurements on L_1 and L_2. The eigenvectors of X is

$$|\text{fail}_X\rangle = \frac{|H\rangle + |T\rangle}{\sqrt{2}}, \quad |\text{ok}_X\rangle = \frac{|H\rangle - |T\rangle}{\sqrt{2}}$$

The eigenvectors of Y is

$$|\text{fail}_Y\rangle = \frac{|\Downarrow\rangle + |\Uparrow\rangle}{\sqrt{2}}, \quad |\text{ok}_Y\rangle = \frac{|\Downarrow\rangle - |\Uparrow\rangle}{\sqrt{2}}$$

The expected contradiction can be derived as follows:

Through the first path: The probability of both W_1 and W_2 measuring "ok" simultaneously is calculated directly as $\frac{1}{12}$, since $|\Psi\rangle = \frac{1}{\sqrt{3}}|H\rangle|\Downarrow\rangle + \sqrt{\frac{2}{3}}|T\rangle|\Rightarrow\rangle$ can be expressed as

$$|\Psi\rangle = \frac{|\text{ok}_X\rangle|\text{ok}_Y\rangle}{\sqrt{12}} - \frac{|\text{ok}_X\rangle|\text{fail}_Y\rangle}{\sqrt{12}} + \frac{|\text{fail}_X\rangle|\text{ok}_Y\rangle}{\sqrt{12}} + \sqrt{\frac{3}{4}}|\text{fail}_X\rangle|\text{fail}_Y\rangle$$

Through the second path: Let $|\Psi$ be re-expressed as

$$|\Psi\rangle = \sqrt{\frac{2}{3}}|\text{fail}_X\rangle|\Downarrow\rangle + \sqrt{\frac{1}{6}}|\text{fail}_X\rangle|\Uparrow\rangle - \sqrt{\frac{1}{6}}|\text{ok}_X\rangle|\Uparrow\rangle$$

From this expression, it can be inferred that if W_1 obtains $|\text{ok}_X\rangle$ in X measurement on laboratory L_1, she can infer that the outcome of F_2's S_z measurement on the qubit must be $+\frac{1}{2}$. In turn, if F_2 obtains $+\frac{1}{2}$, she can infer that the outcome of F_1's measurement on the quantum coin was $|t\rangle$. And if F_1 obtains $|t\rangle$, she can infer that the outcome of W_2's measurement on L_2 will be $|\text{fail}_Y\rangle$, since $|\Rightarrow\rangle = |\text{fail}_Y\rangle$ as definition. Therefore, via this reasoning chain (based on conditional probability), it is concluded that if W_1 measures $\left|\text{ok}_X\right\rangle$, then W_2 must measure $\left|\text{fail}_Y\right\rangle$, resulting in a joint probability of zero. The two results are contradictory.

Formulated in quantum histories, we call the five steps as $t_1,\ldots,t_5$; at t_1, observer F_1 measures the quantum coin; at t_2, F_1 prepares the qubit; at t_3, F_2 measures the qubit; at t_4, W_1 measures laboratory L_1; at t_5, W_2 measures laboratory L_2. However, we cannot combine the histories

$$\hat{C}\left(\hat{P}_{a_t,b_\uparrow,w_{1\text{ok}},w_{2\text{ok}}}\right)=U_{t_0t_1}\hat{P}_{a_t}U_{t_1t_3}\hat{P}_{b_\uparrow}U_{t_3t_4}\hat{P}_{w_{1\text{ok}}}U_{t_4t_5}\hat{P}_{w_{2\text{ok}}}U_{t_5t_0}$$
$$\hat{C}\left(\hat{P}_{a_h,b_\downarrow,w_{1\text{ok}},w_{2\text{ok}}}\right)=U_{t_0t_1}\hat{P}_{a_h}U_{t_1t_3}\hat{P}_{b_\downarrow}U_{t_3t_4}\hat{P}_{w_{1\text{ok}}}U_{t_4t_5}\hat{P}_{w_{2\text{ok}}}U_{t_5t_0}$$

together into a family meeting the consistency condition, because

$$Tr\left[\hat{C}^\dagger\left(\hat{P}_{a_t,b_\uparrow,w_{1\text{ok}},w_{2\text{ok}}}\right)|\Psi_0\Psi_0|\hat{C}\left(\hat{P}_{a_h,b_\downarrow,w_{1\text{ok}},w_{2\text{ok}}}\right)\right]$$
$$=\Psi_0\left|\hat{C}\left(\hat{P}_{a_h,b_\downarrow,w_{1\text{ok}},w_{2\text{ok}}}\right)\hat{C}^\dagger\left(\hat{P}_{a_t,b_\uparrow,w_{1\text{ok}},w_{2\text{ok}}}\right)\right|\Psi_0=\frac{1}{12}\neq 0$$

Thus, it is not surprising that the reasoning within the family will lead to inconsistent conclusions.

These three well-known cases, respectively, pertain to spatial nonlocality, temporal retrocausality, and Wigner's friend–type phenomena. By employing the quantum histories approach for analysis, it becomes evident that similar macro-classical physical descriptions often implicitly presuppose non-commutativity and incompatibility from the very beginning. Consequently, reasonings constructed upon such foundations inevitably give rise to paradoxes and inconsistencies. Nevertheless, despite the fact that history chain operator analysis can clearly reveal the unnoticed non-commutativity and incompatibility, people still find these cases challenging to accept due to their counterintuitive nature. This difficulty was frankly acknowledged by Griffiths, just as he recently advertises his approach to solving quantum paradox:

> …it is worth remarking that the CH (consistent histories) approach, while it is in certain ways counterintuitive - perhaps necessary, given that modern quantum mechanics was developed precisely because classical intuition failed - is much simpler than the earlier quantum logic of Birkhoff and von Neumann…
>
> (Griffiths, 2023: 12)

Notes

1 Some people also refer to this interpretation as the decoherent histories interpretation, many-histories interpretation, histories formalism of quantum mechanics, generalized quantum mechanics, etc. This can reflect the confusion in the usage of terms such as interpretation, theory, formalism, and approach among literature. The decoherent histories interpretation was the term initially adopted by Hartle and Gell-Mann. Later, they, along with Griffiths and Omnès, admitted that the fundamental contents of the consistent

histories interpretation and the decoherent histories interpretation were substantially the same. In this book, it is mostly referred to as the quantum histories interpretation. When sorting out the content from the literature of Hartle, Halliwell and others, it is also cited as the decoherent histories interpretation.

Hartle and Gell-Mann would prefer a physically realistic criterion for the history family that can reproduce the quasi-classical domain or the real world, while Griffiths seems more willing to relax the standard a bit to accommodate more perspectives for expressing physical reality. The clarification by Hartle/Gell-Mann regarding the relationship between their views and those of Griffiths and Omnès is of some historical value to be cited here (Gell-Mann and Hartle, 1994: 342–343):

> ...Griffiths' ideas were extended by Omnès and a similar formulation was arrived at independently but later by ourselves. As far as we are aware, all formulations of the quantum mechanics of closed systems under serious consideration, including ours, are "consistent history formulations" in the sense of requiring the consistency of a set of probability sum rules for those histories that are assigned probabilities. However, there are different formulations depending on just what probability sum rules are required and the strength of the conditions used to ensure their consistency.... The consistency condition of Griffiths is that $\mathrm{Re}D(\alpha',\alpha)\approx 0$, for $\alpha'\neq\alpha$, if and only if $C_{\alpha'}+C_{\alpha}$ is another chain of projections.... What we called the weak-decoherence condition, $\mathrm{Re}D(\alpha',\alpha)\approx 0$, for $\alpha'\neq\alpha$, is stronger than that of Griffiths. It is the necessary condition if histories that are sums of chains of projections are allowed (corresponding to a rule for coarse-graining that allows arbitrary unions of histories as new histories). Our medium-decoherence condition, $D(\alpha',\alpha)\approx 0$, for $\alpha'\neq\alpha$, is still stronger.... In our work we have sought a notion of decoherence of histories that would capture generally the idea that, in physically interesting situations, relevant for quasi-classical behavior, quantum-mechanical interference between histories vanishes for a reason. That is, we sought to provide a general characterization of the mechanisms of dissipation of phases described by Zeh, Zurek and others, but in a way that would not require an artificial or poorly defined division of the closed system into subsystem and environment.... The differences we have described should not obscure the fact that our work lies within the class of consistent histories formulations of the quantum mechanics of closed systems, but should also not obscure the fact that decoherence of histories, as we have defined it, is not the same as just consistency.

2 When the consistency condition is not guaranteed to be satisfied, the history probability does not have such general additivity, but only has additivity for the event probability at the last moment:

$$p\left(\hat{P}_{a_1}^{t_1},\ldots,\hat{P}_{a_{n-1}}^{t_{n-1}}\right)=\sum_{a_n}p\left(\hat{P}_{a_1}^{t_1},\ldots,\hat{P}_{a_{n-1}}^{t_{n-1}},\hat{P}_{a_n}^{t_n}\right)$$

3 This is because when the consistency condition is not satisfied, the following formula may not hold:

$$\sum_{\alpha}P(\alpha)=P\left(\sum_{\alpha}\alpha\right)$$

This distinction between homogeneous and non-homogeneous histories plays an important role in some selection conditions for differentiating history families, see Chapter 6.

4 In Griffiths' view, giving up the principle of uniqueness does not mean denying the existence of an external objective world independent of us; it merely abandons the mode of thinking and reasoning and the single description of the world in classical physics.

5 This involves a significant issue in the interpretation of quantum mechanics, namely the so-called "Heisenberg's cut": how to clearly demarcate the boundary between the classical and the quantum world, whether in specific circumstances or in a general sense? Halliwell has discussed how the probabilities of crossing incompatible multiple history families can be uniformly assigned. He found that in some cases it can be done uniformly, but in many cases it cannot. From this, a new criterion for measuring "quantumness" can be derived, and conversely, it is therefore a criterion for measuring "classicality" (Halliwell, 2017).

6 Although this book does not deal with the issues of relativistic physics and gravitation, if the quantum consistent histories framework can indeed provide a natural depiction of the quasi-classicality derivation process, then it can in principle give a consistent description of the reduction from quantum gravity to classical gravity (Sun, 2017).

7 However, if what is being examined is not position or momentum but more general physical quantities, the probability expression may no longer hold.

8 In fact, Hartle and Gell-man's starting point was the issue of quantum cosmology and the quantizatoin of space-time (Gell-man and Hartle, 1993; Hartle, 1993). Their discussion on the decoherent histories of space-time evolution in the cosmological context incorporated Feynman's formalism. The discussion here is limited to non-relativistic quantum mechanics, so we will not elaborate on this topic more.

9 It is interesting to mention that, in the 1960s, Wigner also had given advice and assistance to Hartle on his early research into the foundations of quantum mechanics, and encouraged him to publish his work (see Hartle's reminiscences in Hartle, 2005: 81).

4 Ontological and logical issues in histories interpretation of quantum mechanics

This chapter discusses the ontological and logical issues involved in criticisms. The ontology of quantum histories interpretation is based on two core contents. One is the stochastic evolution dynamics, which is similar to standard quantum mechanics; the other is the reasoning principles (especially the single-framework rule) that must be followed if the quantum process is expressed in the histories formalism, which makes the interpretation deviate from previous rules of classical propositional reasoning.

Although in those cases where classical mechanics can be regarded as an approximation of quantum mechanics, the "quantum logic" proposed by histories interpretation is enough to reduce to the classical propositional logic, there are still unresolved challenges regarding the relationship between the new and old logics, the new and old ontologies. In particular, there is a certain anti-realism in the expressions of the founders of the quantum histories interpretation, which causes mutual influence between the characterization of the properties of quantum histories and the logical inference of quantum histories propositions.

The previous chapter introduced the interpretation problem of quantum mechanics and the key points and ideological sources of the quantum histories interpretation. This interpretation and its subsequent development have brought new ideas to the interpretation problem of quantum mechanics. Many of its rather novel expressions on the nature of physical reality and physical propositions are quite different from classical physics and quantum mechanics in standard textbooks. Since its proposal, scholars who advocate other approaches, such as the Copenhagen interpretation, the spontaneous collapse theory, the Everettian interpretation, and Bohm mechanics, have criticized it for having serious problems with its ontological commitment to physical reality. Some think it is far from complete and new physical content must be added to it; some think it has an anti-realism tendency; others think that although the consistent histories interpretation is named as such, it is logically inconsistent itself; in addition, some scholars have raised questions about its claims on the approximate validity of consistency conditions, quantum non-locality, and quantum contextuality.

This chapter focuses on the disputes involving ontological and logical issues. The criticisms on these two aspects are the most acute, and their founders have constantly defended and clarified them. Griffiths himself clearly mentioned (Griffiths,

DOI: 10.4324/9781003662860-4

2014, 2024) when discussing the ontology and logic of the consistent histories interpretation that its ontology is based on two core contents: one is the evolution dynamics with randomness, which is similar to standard quantum mechanics and some other versions of interpretation and has not caused conceptual disputes; the other part is the reasoning principles (especially the single-framework rule) that need to be followed after the quantum physical process is expressed in the form of consistent histories. The series of problems arising here have made the interpretation deviate to a large extent from the classical world picture and the previous rules of proposition reasoning.

Although Griffiths and Omnès et al. believe that in those cases where classical mechanics can be regarded as an approximation of quantum mechanics, the "quantum logic" proposed by the quantum histories interpretation can be reduced to the familiar classical propositional logic, there are still many ambiguous points that have not been clarified in the relationship between the new and old logics and the new and old ontologies. Especially the anti-realism color in the expressions of the founders of the consistent histories interpretation creates a mutual influence between the characterization of the attributes of quantum histories and the logical inference of quantum history propositions. Before examining the quantum histories formalism from the perspective of nomology in the following sections, this chapter first sorts out these two aspects of problems.

4.1 Ontological issues in histories interpretation of quantum mechanics

Although Griffiths and others have used the quantum histories interpretation, especially the single-framework rule, to dissolve many of the puzzles and paradoxes in quantum mechanics, it is also believed that a considerable ontological "cost" has been paid in this process. Many scholars have raised questions about the ontological issues involved (such as Dowker and Kent, 1996; Bassi and Ghirardi, 1999; Stapp, 1999, Stapp, 2012; Okon and Sudarsky, 2014a, 2014b, 2015). Losada and Lombardi systematically organized and classified the academic community's doubts and criticisms of the quantum histories interpretation over the years, dividing them into theoretical and interpretative aspects (Losada and Lombardi, 2018). Referring to the introduction in Chapter 2, the so-called "theoretical aspect" and "interpretative aspect" can roughly be understood as: the formalism of quantum histories can be regarded as a theoretical extension without altering the fundamental mathematical structure of quantum mechanics, and it does not directly come with a ready-made interpretation; while the interpretative aspect focuses on clarifying the ontological elements within it (see Section 2.2 of Chapter 2). However, their classification of various criticisms does not strictly follow the distinction they made between theory and interpretation. Here, we will reorganize these contents from the ontological and logical perspectives. This section first examines the ontological doubts and criticisms.

4.1.1 Discontinuity in the ascription of properties

Some have pointed out that the consistency conditions proposed by Griffiths are overly restrictive. This is mainly reflected in that within a consistent history family, the physical properties of a system may undergo discontinuous changes within arbitrarily close adjacent moments, and even switch to non-commuting properties within an arbitrarily short period of time, which violates the basic constraints on the system's state imposed by the uncertainty principle of quantum mechanics. Taking a spin system as an example, consider a very simple history family (Laura and Vanni, 2009: 164):

At the initial moment t_0, the system is in the pure state $\rho_0 = |+x\rangle\langle +x|$, and the Hamiltonian is always 0; at the subsequent moment t_1, the spin of the system is in the x-direction, with possible situations being $\hat{P}_{t_1 x\uparrow} = |+x\rangle\langle +x|$ or $\hat{P}_{t_1 x\downarrow} = |-x\rangle\langle -x|$; at the next moment t_2, the spin of the system is in the z-direction, with possible scenarios being $\hat{P}_{t_2 z\uparrow} = |+z\rangle\langle +z|$ or $\hat{P}_{t_2 z\downarrow} = |-z\rangle\langle -z|$. It can be verified that this history family is consistent, and thus the probabilities of each history can be calculated, such as

$$Pr\left(\hat{P}_{t_1 x\uparrow}, \hat{P}_{t_2 z\uparrow}\right) = Tr\left(\hat{P}_{t_2 z\uparrow} \hat{P}_{t_1 x\uparrow} \rho_0 \hat{P}_{t_1 x\uparrow} \hat{P}_{t_2 z\uparrow}\right) = \left|\langle +x|+z\rangle\right|^2 = \frac{1}{2}$$

However, the problem is that there are no restrictions on the selection of t_1 and t_2. Therefore, the above quantum history analysis cannot rule out the situation where the two are arbitrarily close:

$$\lim_{t_1 \to t_2} Pr\left(\hat{P}_{t_1 x\uparrow}, \hat{P}_{t_2 z\uparrow}\right) = \frac{1}{2}$$

When $t_1 = t_2$, it clearly violates the requirement of the uncertainty principle that spins in different directions cannot have definite values at the same time. Similar constructions can be easily given. This indicates that the quantum histories interpretation cannot avoid such situations.

4.1.2 Difficulty in maintaining quasi-classicality

We live in a classical world, which is obviously persistent and roughly stable at the macroscopic scale. However, the quantum histories formulation seems unable to guarantee this. As introduced in Chapter 3, the founders of the consistent histories interpretation proposed a set of explanations for the quasi-classicality problem. However, from the previous objection, it can be inferred that the so-called quasi-classicality generally cannot be maintained in the consistent history family: the history belonging to a certain quasi-classical domain can only maintain quasi-classicality for a certain period of time, and it is not guaranteed that it can be maintained at future moments. Even if we "consistently" extend a history family to future time points, such an extension may not always maintain quasi-classicality.

Since the supporters of the quantum histories interpretation claim to dissolve the measurement problem and wave function collapse, and the measurement process itself is undoubtedly a quasi-classical process involving macroscopic properties, the problem that quasi-classicality cannot always be maintained is detrimental to the persuasiveness of the quantum histories interpretation. Dowker and Kent pointed out (Dowker and Kent, 1996: 1634):

> ...persistent quasi-classicality is such a basic feature of our experience, and failure to predict it leaves such a significant gap in our understanding of the world, that we feel that some argument, or some new interpretation, or some new formalism, that would amend the situation should be sought.

Here, a theory refers to a specific framework selection method for the process of changes in system properties.

Griffiths' response to this (Griffiths, 2013: 104) is that although the situation criticized by Kent and others can indeed occur in the quantum histories interpretation, this interpretation does not have to guarantee the persistence of quasi-classicality – because it is never a property directly possessed by "this world", but only a property of "our description of the world", so there is no conceptual difficulty here. Clearly, this response still reflects Griffiths' unchanged descriptivism and anti-realism stance toward the quantum histories interpretation, which seems undesirable to realists.[1]

4.1.3 Realm-related reality

In the quantum histories interpretation, which physical processes are real depends on the selection of the consistent history family (some literature refers to it as "realm") they belong to. In fact, according to the quantum histories interpretation, no selection method is special or prioritized; all methods are equal (via the principle of equality); in addition, the principle of liberty allows physicists to freely choose any suitable history family for analysis based on practical needs. Okon and Sudarsky criticized that this degree of freedom means that the actual physical processes and the physical properties of the system will also depend on the choice of the history family. Reality becomes the relativized thing that depends on the specific history family chosen in each case (Okon and Sudarsky, 2014a: 27–30). This is contrary to the realist picture.

Griffiths' response to such criticism is (Griffiths, 2013: 99): selecting different realms in the quantum histories interpretation is similar to selecting different inertial frames in the framework of special relativity. In special relativity, there is no preferred inertial frame, and all inertial frames are on an equal footing in describing physical processes, but obviously, not all inertial frames are convenient for describing all physical processes. Okon and others disagree with this inaccurate analogy. In special relativity, all things in all inertial frames can be described in a unified, reference frame-independent, non-relativized formulation, and thus are not relativized realities. But the quantum history analysis clearly cannot do this; here, it is not allowed to combine or unify incompatible history families. This relativity cannot be simply compared with that "relativity" in the relativity theory of space-time. Griffiths seems to admit that this analogy may not be accurate enough, but he does not offer more justification. The coarse-graining and fine-graining methods emphasized by Hartle and Gell-Mann also have similar arbitrariness. Physicists surely hope to find an optimal "coarse/fine-graining" to describe the system, that is, if further fine-graining would violate the consistency condition, and further coarse-graining would lose some descriptive ability without increasing the consistency. But how can we find this optimal degree in a given scenario? In fact, this arbitrariness comes from the arbitrariness of the selection of observables (and this arbitrariness of the selection of observables precisely comes from the arbitrariness of the selection of history families), so there is no objective best standard for coarse/fine-graining.

4.1.4 The dependence of initial states and consistency conditions on the framework

The second criticism of Okon et al. is that the quantum histories interpretation does not clearly give the relationship between the initial state of the system and the system's Hamiltonian. The theory claims that the initial state ρ and the Hamiltonian H of the system and the time translation operator $U = e^{-iH(t_0 - t_i)/\hbar}$ are independent of the framework selection. However, the framework dependence of the physical process description in the quantum histories interpretation makes it unclear how to obtain this framework-independent initial state and Hamiltonian information within the theory. This is an extension of the previous doubts.

Hartle and Gell-Mann proposed early on that these two aspects need to be given through external information and cannot be directly provided within the interpretation (Gell-Mann and Hartle, 1990: 425–458). But this still cannot solve this conceptual difficulty, because if even "what is real" depends on the framework, then how can our observations of the initial state and the Hamiltonian evolution of the system remain independent and be used as external information?

Furthermore, as will be pointed out in the next section, the past state of the system can only be retraced with the help of the framework. Then, as an earlier state in a history, how can the "initial state" be independent of the framework and remain the same in different frameworks? In addition, although in standard quantum mechanics, we have a way in principle to determine the initial state and Hamiltonian of the system, etc., in the framework-dependent quantum histories system, we cannot simply transplant the methods of standard quantum mechanics to obtain this information.

Moreover, this initial state dependence brings other problems to the ontology of the quantum histories interpretation. As is well known, in the usual quantum mechanics formulation, the quantum state is defined first and has a theoretical priority over observables; in the C^*-algebra formulation of quantum mechanics (sometimes called "algebraic quantum mechanics"), once the C^*-algebra formed by all observables (that is, all possible observable properties of the system) is determined, the quantum state of the system can be defined as some functionals on this algebra, mapping each observable to the complex number field (see Chapter 2). Here, the state is attached to the observable and is only a mathematical tool for calculating probabilities, while the observable is more fundamental. But in the quantum histories interpretation, the basic element, the history with temporal sequence and multiple moments, not only depends on the framework but also on the selection of the initial state. Then, does the initial state have the same (at least not subordinate to the history in logic) ontological status as histories? If so, then the ontological presuppositions of the quantum histories system would seem more complex than those of standard quantum mechanics and algebraic quantum mechanics, and it seems that there is multiple levels of reality, bringing more confusion to the interpretation.

4.1.5 The measurement problem has not been thoroughly solved

The essence of the measurement problem in standard quantum mechanics is that there are two fundamentally different physical processes, one is the deterministic process of the system's wave function evolving according to the Schrödinger equation, and the other is the non-deterministic, random collapse process of the wave function that does not satisfy the Schrödinger equation. To depict the measurement process, it is necessary to know in advance which part is the process that satisfies the Schrödinger equation and which part is the collapse process. The basis for this distinction is beyond the standard quantum mechanics theory and is determined by external factors. The quantum histories interpretation always emphasizes that it can dissolve this problem because there is no need for the wave packet collapse assumption to describe the measurement process. However, Okon and Sudarsky's doubts about this are (Okon and Sudarsky, 2014b: 10–11):

1 Apart from the principle of utility, this interpretation does not propose a method for choosing the domain in specific situations. Even if there is a way to filter out such a domain in the measurement process, this filtering rule is external to the interpretation.
2 Even if a domain is determined, there is no way to determine the status of different histories within it. If only one history is actual, then this interpretation is incomplete because it fails to provide an explanation for the particularity of this history; if all histories are real and realized, then the values assigned to each branch by the Born rule in the context of the domain cannot be understood as probabilities, which also inherits similar difficulties faced by the many-worlds interpretation. This involves the view of the quantum histories interpretation as a "many-histories interpretation" and its inheritance relationship with the traditional many-worlds interpretation. Swiss mathematical physicist Fröhlich et al. also pointed out that for two incompatible multi-time event sequences $\{a_1;\ldots;a_n\}$ and $\{b_1;\ldots;b_m\}$, it is difficult to determine which of the two occurs. If we assume that both occur, it would be similar to the many-worlds picture (Fröhlich and Schubnel, 2013: 26). Therefore, transplanting the defense of the

probability problem by the supporters of the many-worlds theory to the defense of the quantum histories interpretation might be a feasible approach (for the discussion of the many-worlds theory, see the appendix of Chapter 3).

4.1.6 Lack of specific explanations for the quasi-classicality experience in the macroscopic world

As mentioned in Chapter 3, Hartle and Gell-Mann proposed the concept of information gathering and utilization systems (IGUSes) in order to uniformly apply the quantum histories interpretation to comprehensively explain all phenomena in the microscopic and macroscopic worlds, especially to explain why human beings and other such subjects only experience this quasi-classical world and can discover and utilize the quasi-classical regularities to survive in it. IGUS can adapt to the environment independently and can take advantage of the predictability in the quasi-classical domain (that is, discover the rules within it), especially to utilize the physical laws within it to adapt to its environment. Hartle asserts that the reason why we have evolved to the point where we can make predictions in the quasi-classical world is that the quasi-classical world has enough simple regularities, and our evolution has adapted to the environment. Okon and Sudarsky's objection is that the quantum histories interpretation itself does not contain any description of such macroscopic evolutionary processes. The four essential elements for evolution – diverse initial populations, external environment, genetics, and selection – are all absent here. Especially the element of "external environment", it contradicts Hartle's previous assertion, because the quasi-classical domain itself is understood as the product and result of "adaptive selection", and it cannot simultaneously be the external environment that causes the selection to occur (Okon and Sudarsky, 2014a: 31).

Brief summary of the above content: The main controversial points of the ontology given by the quantum histories interpretation are that all physical realities depend on the framework and the initial state. Griffiths even claims that the quantum history analysis in each framework merely aims to provide a partial description of reality, which leads to the possibility that the theoretical presuppositions it gives may slide toward relativism. Regarding the measurement problem, which is a difficult interpretative issue in quantum mechanics, the solution provided by the quantum histories interpretation, although it formally eliminates the dual dynamics mechanism of measurement, is either considered incomplete because it cannot pick out the only truly occurring history, or it has to admit the reality of all histories like the multi-world interpretation, and also faces the problem of understanding probability. The quasi-classical world where humans exist can be derived from the expression of this system, although it has some explanatory power, but this process is currently only a vision and cannot be thoroughly explained within the interpretation.

4.2 Logical issues in histories interpretation of quantum mechanics

Griffiths once specifically discussed the logical problems in the quantum histories interpretation and expounded the differences between it and the so-called "quantum logic" (Griffiths, 2014). The logical reasoning of propositions in the quantum histories interpretation is different from what many quantum logic schemes have attempted to do, which is to find an alternative logical axiom system and orthomodular lattice algebra different from classical logic and Boolean algebra to eliminate paradoxes. The logic Griffiths mentioned mainly refers to the four reasoning principles, especially the single-framework rule – by adhering to them, it can be guaranteed that all event proposition reasoning within the same framework satisfies classical logic and all probability operations conform to the Boolean operations of classical probability propositions. However, this approach has not been accepted by all scholars. The main doubts about the logical aspect of the quantum histories interpretation are of three types: the first is that contradictory results may occur when making predictions and retrodictions from known consistent history families; the second is that probability proposition reasoning across history families may lead to contradictory conclusions; the third is that when involving contextuality correlations of observable quantities, even the truth value assignment of the same proposition in two compatible frameworks may be contradictory. The following examples will illustrate these three situations.

4.2.1 *Contradictory retrodictions and predictions*

Retrodiction refers to, given the physical properties of the system at the present moment and the dynamical equation, to infer what the physical properties were at an earlier time. In classical physics, the result of such retrodiction is generally definite and unique, and thus is not contradictory. However, in the quantum histories interpretation, the description of the physical system's properties, as well as the properties themselves, depend on the selection of consistent history families. Therefore, a property that the system has at the present moment may be retrodicted to be mutually contradictory properties at an earlier time in two different consistent history families (and thus cannot have both actually occurred).

Specifically, there may be several relations defined between two projection operators $\hat{P}_a$, $\hat{P}_b$ representing different properties of the system[2]:

Complementary (non-commutative) relation: $\hat{P}_a\hat{P}_b \neq \hat{P}_b\hat{P}_a$;

Contradictory relation: $\hat{P}_a + \hat{P}_b = \hat{I}$, where $\hat{I}$ is the identity operator;

Contrary relation: $\hat{P}_a \perp \hat{P}_b$, but $\hat{P}_a + \hat{P}_b \neq \hat{I}$.

In standard quantum mechanics, given two contrary properties p, q of a system and the system's state ρ, and represented by the corresponding projection operators, there can never exist a system property r such that the conditional probabilities of p and q relative to r are both 1 under state ρ. However, in the quantum histories interpretation, Kent gave an example to point out that the retrodiction of the probability of an attribute's occurrence must depend on the selection of the framework (Kent, 1997). The quantum histories interpretation allows us to retrodict the properties of the system that are in a contrary relationship in different frameworks after giving the initial state of the system, and the probabilities of their occurrence are not zero. In other words, here, given the current data information, we generally cannot uniquely infer the state and properties of the system in the past. Griffiths and Hartle's response to this is that as long as a unique history state can be retrodicted in each selected consistent history family, the physical events occurring in incompatible history families cannot be examined simultaneously (Griffiths and Hartle, 1998). Despite this, it is questionable that Hartle et al. repeatedly emphasize that a major theoretical advantage of the quantum histories interpretation is its applicability to the description of closed systems, including the entire universe. One of the main contents of quantum cosmology research is to trace back to the initial state of the universe at the Big Bang. If it lacks the ability to trace back and determine the (unique and definite) initial state, it will be difficult to achieve the goal of tracing the evolution process of the universe.

In addition to the possible inconsistent results in the retrograde analysis of the past, the quantum histories interpretation also faces the problem of lacking predictive power for future events/attributes of physical systems. Since any future attribute can only belong to a selected possible framework, and any choice is equivalent in principle, in the same initial state and different history families, there may be a situation where the probability of mutually orthogonal histories occurring is 1. In other words, the theory predicts that two opposing physical attributes will "necessarily" occur, which is quite different from people's usual understanding of physical processes, making the reasoning and judgment about what attributes a physical system will have at a certain future time meaningless (Dowker and Kent, 1996; Halliwell, 2017: 8). This prediction is not just simply due to the inherent probability of quantum mechanics and the inability to determine which branch will occur. Here, even "what events/attributes are possible" depends on arbitrary human choices.

The principle of utility proposed by Griffiths states that in each specific physical situation, some selection methods of history families are more effective and suitable for describing physical systems than others. Therefore, for a certain experimental measurement process, we can always select a history family with good effectiveness, within which a definite prediction of future attributes can be made. The problem is that the quantum histories interpretation starts with the elimination of wave function collapse as its starting point, so it cannot distinguish measurement processes from non-measurement processes in principle within the interpretation. The so-called effectiveness measurement is also quite arbitrary, unless it returns to the old discussion of the quantum measurement problem and wave function collapse problem.

4.2.2. *Problems in the reasoning process of probability propositions across history families*

Let's illustrate with an example. First, define the following seven vectors:

$$|a\rangle = \frac{1}{\sqrt{5}}\begin{pmatrix}1\\1\\1\\1\\1\end{pmatrix}$$

$$|c\rangle = \frac{1}{\sqrt{5}}\begin{pmatrix}1\\1\\1\\-1\\-1\end{pmatrix}$$

$$|b_1\rangle = \begin{pmatrix}1\\0\\0\\0\\0\end{pmatrix}, |b_2\rangle = \begin{pmatrix}0\\1\\0\\0\\0\end{pmatrix}, |b_3\rangle = \begin{pmatrix}0\\0\\1\\0\\0\end{pmatrix}, |b_4\rangle = \begin{pmatrix}0\\0\\0\\1\\0\end{pmatrix}, |b_5\rangle = \begin{pmatrix}0\\0\\0\\0\\1\end{pmatrix}$$

From them, seven projection operators can be, respectively, constructed, and the history family can be constructed based on this. Consider the following three history families each containing three moments:

$$\mathcal{F}_1 : \left\{ \begin{matrix} |a\rangle\langle a| \\ I - |a\rangle\langle a| \end{matrix} \right\}, \left\{ \begin{matrix} |b_1\rangle\langle b_1| \\ I - |b_1\rangle\langle b_1| \end{matrix} \right\}, \left\{ \begin{matrix} |c\rangle\langle c| \\ I - |c\rangle\langle c| \end{matrix} \right\}$$

$$\mathcal{F}_2 : \left\{ \begin{matrix} |a\rangle\langle a| \\ I - |a\rangle\langle a| \end{matrix} \right\}, \left\{ \begin{matrix} |b_1\rangle\langle b_1| + |b_4\rangle\langle b_4| \\ I - |b_1\rangle\langle b_1| + |b_4\rangle\langle b_4| \end{matrix} \right\}, \left\{ \begin{matrix} |c\rangle\langle c| \\ I - |c\rangle\langle c| \end{matrix} \right\}$$

$$\mathcal{F}_3 : \left\{ \begin{matrix} |a\rangle\langle a| \\ I - |a\rangle\langle a| \end{matrix} \right\}, \left\{ \begin{matrix} |b_1\rangle\langle b_1| + |b_5\rangle\langle b_5| \\ I - |b_1\rangle\langle b_1| - |b_5\rangle\langle b_5| \end{matrix} \right\}, \left\{ \begin{matrix} |c\rangle\langle c| \\ I - |c\rangle\langle c| \end{matrix} \right\}$$

From the calculation, we see that $\mathcal{F}_1$, $\mathcal{F}_2$, and $\mathcal{F}_3$ all meet the consistency condition. Consider the following four histories:

$$H = |a\rangle\langle a| \otimes I \otimes |c\rangle\langle c|$$

$$B_1 = |a\rangle\langle a| \otimes |b_1\rangle\langle b_1| \otimes |c\rangle\langle c|$$

$$B_2 = |a\rangle\langle a| \otimes \left(|b_1\rangle\langle b_1| + |b_4\rangle\langle b_4|\right) \otimes |c\rangle\langle c|$$

$$B_3 = |a\rangle\langle a| \otimes \left(|b_1\rangle\langle b_1| + |b_5\rangle\langle b_5|\right) \otimes |c\rangle\langle c|$$

Calculate the four corresponding chain operators for the above history:

$$\hat{H} = \frac{1}{5} | a\rangle\langle c|$$
$$\hat{B}_1 = \frac{1}{5} | a\rangle\langle c|$$
$$\hat{B}_2 = 0, \ \hat{B}_3 = 0$$

It is observed that the history H exists in $\mathcal{F}_1$, $\mathcal{F}_2$ and $\mathcal{F}_3$. The following proposition inferences are respectively examined in $\mathcal{F}_1$, $\mathcal{F}_2$ and $\mathcal{F}_3$:

$$\left(\hat{H} \Rightarrow \hat{B}_1\right) = I$$
$$\left(\hat{H} \Rightarrow \hat{B}_2\right) = 0$$
$$\left(\hat{H} \Rightarrow \hat{B}_3\right) = 0$$

Three probability propositions can be deduced then:

$$P\left(\hat{H} \Rightarrow \hat{B}_1\right) = 1 \quad \text{in } \mathcal{F}_1$$
$$P\left(\hat{H} \Rightarrow \hat{B}_2\right) = 0 \quad \text{in } \mathcal{F}_2$$
$$P\left(\hat{H} \Rightarrow \hat{B}_3\right) = 0 \quad \text{in } \mathcal{F}_3$$

However, if A and B denote any two arbitrary propositions, then in accordance with the rules of probabilistic propositional reasoning, it should hold that:

$$\left(P(A) = 1\right) \wedge \left(P(B) = 1\right) \Rightarrow P(A \wedge B) = 1$$

as well as

$$\left(P(A) = 1\right) \vee \left(P(B) = 1\right) \Rightarrow P(A \vee B) = 1$$

But

$$\left(\hat{H} \Rightarrow \hat{B}_2\right) \wedge \left(\hat{H} \Rightarrow \hat{B}_3\right) = \left(\hat{H} \Rightarrow \hat{B}_1\right)$$
$$\left(\hat{H} \Rightarrow \hat{B}_2\right) \vee \left(\hat{H} \Rightarrow \hat{B}_1\right) = \left(\hat{H} \Rightarrow \hat{B}_2\right)$$

These two formulas, respectively, contradict the above two rules. This indicates that in the quantum histories interpretation, the probability proposition reasoning across history families will lead to conclusions that obviously violate classical logic (Priebe, 2012: 51–56).

Apart from the problem of the contradiction in the results of proposition reasoning, the quantum histories interpretation must also confront an important theorem involving contextuality in quantum mechanics, namely the Kochen–Specker (K–S) theorem (Kochen and Specker, 1967). This theorem focuses on whether the truth value assignment of a certain property proposition in a quantum system can be independent of the truth value assignment of other property propositions without changing some obviously valid functional relations between propositions intuitively. They demonstrated that when the dimension of the Hilbert space is no less than 3, such a situation may occur. There are now many variants and simplified versions of the original proof by Kochen and Specker. Currently, various arguments have utilized this theorem to question the quantum histories interpretation. Here, we analyze one of the stronger versions.

First, introduce the following version of the K–S theorem (Cabello et al., 1996). Consider a set of projection operators $\{P_1, P_2, \ldots, P_{18}\}$ (without normalization) in a four-dimensional Hilbert space:

$$P_1 = \begin{pmatrix} 0 \\ 0 \\ 0 \\ 1 \end{pmatrix}, P_2 = \begin{pmatrix} 0 \\ 0 \\ 1 \\ 0 \end{pmatrix}, P_3 = \begin{pmatrix} 1 \\ 1 \\ 0 \\ 0 \end{pmatrix},$$

$$P_4 = \begin{pmatrix} 1 \\ -1 \\ 0 \\ 0 \end{pmatrix}, P_5 = \begin{pmatrix} 0 \\ 1 \\ 0 \\ 0 \end{pmatrix}, P_6 = \begin{pmatrix} 1 \\ 0 \\ 1 \\ 0 \end{pmatrix}$$

$$P_7 = \begin{pmatrix} 1 \\ 0 \\ -1 \\ 0 \end{pmatrix}, P_8 = \begin{pmatrix} 1 \\ -1 \\ 1 \\ -1 \end{pmatrix}, P_9 = \begin{pmatrix} 1 \\ -1 \\ -1 \\ 1 \end{pmatrix},$$

$$P_{10} = \begin{pmatrix} 0 \\ 0 \\ 1 \\ 1 \end{pmatrix}, P_{11} = \begin{pmatrix} 1 \\ 1 \\ 1 \\ 1 \end{pmatrix}, P_{12} = \begin{pmatrix} 0 \\ 1 \\ 0 \\ -1 \end{pmatrix}$$

$$P_{13} = \begin{pmatrix} 1 \\ 0 \\ 0 \\ 1 \end{pmatrix}, P_{14} = \begin{pmatrix} 1 \\ 0 \\ 0 \\ -1 \end{pmatrix}, P_{15} = \begin{pmatrix} 0 \\ 1 \\ -1 \\ 0 \end{pmatrix},$$

$$P_{16} = \begin{pmatrix} 1 \\ 1 \\ -1 \\ 1 \end{pmatrix}, P_{17} = \begin{pmatrix} 1 \\ 1 \\ 1 \\ -1 \end{pmatrix}, P_{18} = \begin{pmatrix} -1 \\ 1 \\ 1 \\ 1 \end{pmatrix}$$

The above operators can be interpreted as the spin states of a system composed of two electrons $\left((a,b)_1 \otimes (c,d)_2\right)$ in the Pauli matrix representation. For instance, the proposition represented by P_4 is "the first electron's spin is up along the z-axis and the second electron's spin is down along the x-axis". Specifically, after obtaining the expression of the tensor product using the standard notation of Pauli matrices, apply them to these vectors and examine which Pauli operator's eigenvectors they are. They precisely make each column of the four operators a decomposition of the identity I:

D_1	D_2	D_3	D_4	D_5	D_6	D_7	D_8	D_9
P_1	P_1	P_8	P_8	P_2	P_9	P_{16}	P_{16}	P_{17}
P_2	P_5	P_9	P_{11}	P_5	P_{11}	P_{17}	P_{18}	P_{18}
P_3	P_6	P_3	P_7	P_{13}	P_{14}	P_4	P_6	P_{13}
P_4	P_7	P_{10}	P_{12}	P_{14}	P_{15}	P_{10}	P_{12}	P_{15}

The functional v that assigns values to these quantum propositions satisfies the following equations:

$$v(P_1)+v(P_2)+v(P_3)+v(P_4)=1 \qquad v(P_1)+v(P_5)+v(P_6)+v(P_7)=1$$
$$v(P_8)+v(P_9)+v(P_3)+v(P_{10})=1 \qquad v(P_8)+v(P_{11})+v(P_7)+v(P_{12})=1$$
$$v(P_2)+v(P_5)+v(P_{13})+v(P_{14})=1 \qquad v(P_9)+v(P_{11})+v(P_{14})+v(P_{15})=1$$
$$v(P_{16})+v(P_{17})+v(P_4)+v(P_{10})=1 \qquad v(P_{16})+v(P_{18})+v(P_6)+v(P_{12})=1$$
$$v(P_{17})+v(P_{18})+v(P_{13})+v(P_{15})=1$$

It is observed that each of the 18 operators appears exactly twice. Therefore, any functional that assigns values to this set of propositions will necessarily assign an even number of terms (the propositions within) as true, meaning the sum of all terms on the left side is an even number. However, the total number of columns in the table is odd, and the sum on the right side is odd, which leads to a contradiction. Thus, there does not exist a uniform truth value assignment for these 18 propositions.

Griffiths' single-framework rule can explain the above contradiction as follows: the truth value functional v defined on $D_1 \cup \cdots \cup D_9$ cannot be discussed within the same consistent history family. It encompasses many non-commuting physical properties (such as the electron spin in two perpendicular directions), which results in the non-existence of a consistent history family that can accommodate all of them.

Similarly, for the Peres–Mermin and Kochen–Specker paradoxes, Griffiths points out: the key to identifying the source of the paradox is the observation that though operators in a particular row commute with each other, and hence can be simultaneously assigned values, however, pairs of operators that are not in the same row or the same column are incompatible. They just do not commute. Consequently, the argument leading to the paradox does not satisfy the single-framework rule of quantum histories interpretation: it violates a fundamental principle of quantum reasoning (Griffiths, 2024, Section 8.2.2).

Thus, the truth value functional defined in this way in the quantum histories interpretation is meaningless. According to Griffiths' principle, the truth value functional must be interpreted as "reflecting the system state represented by the history propositions defined in the same consistent history family".

In earlier literature (Bassi and Ghirardi, 1999: 258–260), Bassi A. and Ghirardi G. used the K–S theorem to argue that if each consistent history family is equipped with a truth value assignment, that is, a homomorphism F from a Boolean set of history propositions to the truth value $\mathbb{Z}_2 := \{0,1\}$, then a history proposition α may be assigned the truth value $F(\alpha)=1$ in one history family, but the history proposition α' representing the same event in another history family that may not be compatible with it may no longer have $F'(\alpha')=1$. In this way, the truth or falsehood of the proposition will be erratic. Griffiths still used the single-framework rule to respond to this objection, arguing that the inconsistency of the truth value of the same history proposition across frameworks does not imply the inconsistency of the quantum histories interpretation itself. However, Priebe later argued that using the K–S theorem can lead to an even more counter-intuitive conclusion: if such a Boolean assignment mapping exists, then even within two compatible frameworks, assigning the same truth value to the same history proposition will lead to a contradiction (Priebe, 2012: 49–50). We will briefly recapitulate the proof below. Note that the following elaborations are compiled from Priebe's original proof without substantial modification. Readers can refer to his original text for more details.

Consider the nine decompositions of the identity operator using 18 projection operators $\{D_1,\ldots,D_9\}$ mentioned above, with the subscripts increasing from small to large representing the time sequence from earlier to later.[3] First, let the history family corresponding to each identity operator decomposition D_i be denoted as $\mathcal{F}_i$, and the history family formed by two identity operator decompositions $\{D_i, D_j\}$ be denoted as $\mathcal{F}_{ij}$. To simplify the argument, only the case where all initial states are the maximally mixed state is considered. Let the Boolean algebra formed by the operations between history propositions in the history family be denoted as $\mathcal{B}(\mathcal{F}_i)$ and $\mathcal{B}(\mathcal{F}_{ij})$, and the truth value functional on this Boolean algebra be denoted as θ_i and θ_{ij}. It can be proved that there exists at least one pair $\{D_i, D_j\}$ and $P \in D_i \cap D_j$ such that one of the following two equations

$$\theta_i(P) = \theta_{ij}(P \otimes I)$$

$$\theta_j(P) = \theta_{ij}(I \otimes P)$$

does not hold. Define a truth value function $\Theta : D_1 \cup \cdots \cup D_9 \mapsto \{0,1\}$ as

$$\Theta(P) = \theta_i(P) \quad \text{when } P \in D_i$$

First, the well-definedness of Θ should be clarified, namely there will be no case that "for a certain P, the value of $\Theta(P)$ cannot be determined". Noticc that $P \in D_i \cap D_j$, and then it follows that

$$\theta_{ij}\left((P \otimes I) \rightarrow (I \otimes P)\right) = \theta_{ij}\left(\overline{P \otimes (I - P)}\right) = 1$$

Here, since $P \otimes (I - P)$ is "dynamically impossible", the occurrence probability of the history containing such an event is 0, $P \otimes (I - P)$ is false, that is, $\overline{P \otimes (I - P)}$ is true. From the above equation, we can obtain

$$\begin{aligned}
\theta_{ij}(P \otimes I) &= \theta_{ij}(P \otimes I) \cdot 1 = \theta_{ij}(P \otimes I) \cdot \theta_{ij}\left((P \otimes I) \rightarrow (I \otimes P)\right) \\
&= \theta_{ij}\left((P \otimes I) \wedge \left((P \otimes I) \rightarrow (I \otimes P)\right)\right) \\
&= \theta_{ij}\left(\left((P \otimes I) \wedge \left(\overline{P \otimes I}\right)\right) \vee \left((P \otimes I) \wedge (I \otimes P)\right)\right) \\
&= \theta_{ij}\left((P \otimes I) \wedge (I \otimes P)\right) = \theta_{ij}(P \otimes P)
\end{aligned}$$

Similarly, there is $\theta_{ij}(I \otimes P) = \theta_{ij}(P \otimes P)$. Thus, we obtain

$$\theta_{ij}(P \otimes I) = \theta_{ij}(I \otimes P)$$

This means that

$$\theta_i(P) = \theta_{ij}(P \otimes I) = \theta_{ij}(I \otimes P) = \theta_j(P)$$

Therefore, even if P belongs to both D_i and D_j simultaneously, Θ is well-defined, then the value of $\Theta(P)$ is unique.

Since Θ is non-contextual, only one proposition in each D_i is assigned the value of true, the above K-S paradox will occur again: each operator appears twice, but the total number of identity decompositions is odd. From this, it can be known that there exists a pair of i and j, either $\theta_i(P) = \theta_{ij}(P \otimes I)$ or $\theta_j(P) = \theta_{ij}(I \otimes P)$ is false. Suppose the former is false, which means that for the consistent history family $\mathcal{F}_i$ and its fine-graining $\mathcal{F}_{ij}$, the truth value of P in $\mathcal{F}_i$ and the truth value of $P \otimes I$ in $\mathcal{F}_{ij}$ are actually different. Note that $\mathcal{F}_i$ and $\mathcal{F}_{ij}$ are obviously two compatible history families. This result clearly violates intuition and does not satisfy classical logic.

Priebe gives a further inference: in a four-dimensional Hilbert space, there does not exist such a truth functional that can keep the assignment of the value of any history proposition in a consistent history family consistent with the assignment of the same proposition in the fine-graining result of that history family.

4.3 Additional remarks of the ontological and logical issues

In summary, the ontological doubts about the quantum histories interpretation mainly focus on the fact that its description of the attributes, events, and processes of physical systems is highly dependent on the selection of the framework/realm. Even the quasi-classical realm that its supporters expect to find is no exception, because the consistent/decoherent conditions alone are far from guaranteeing the persistence of the quasi-classical realm. The dependency of the consistency determination and probability calculation of the history family on the initial state and Hamiltonian information input from outside the history family further weakens the independence of the theoretical ontology and may also lead to the complication of the ontological hierarchy.

The criticism on the logical aspect indicates that it is impossible to reliably judge the truth or falsehood of the attribute propositions outside the time support of the already determined consistent history family based on the propositions within it, nor can it be inferred whether the same history proposition appearing in other history families, whether compatible or not with the previous family, is true or false.

It should be noted that there is a significant correlation between the inference of history propositions and the assignment of properties to the system in the quantum histories interpretation. Many ontological issues also reflect the underlying problem of history proposition reasoning, and the latter is also related to the former. As mentioned at the beginning of this chapter, Griffiths himself has always believed that the ontological basis of the quantum histories interpretation cannot be separated from the principle of proposition reasoning within a single framework. From the perspective of the "outside" of the theory, the classical physical thinking in the past holds that the truth value of all physical propositions has objectivity that is independent of specific expressions, and we can reliably infer the content connection between propositions from the truth value relation among multiple propositions. However, in the quantum histories interpretation, since history events all depend on the framework, all propositions expressing history events cannot be stripped out of this framework-dependent information for examination.

Notes

1 Omnès (2004: 255) held a similar attitude: "The language of consistent histories interpretation cannot pretend to be a realistic description of micro-events; … The real limitation does not lie in the actual existence, but in our language".
2 Simply speaking, due to the limited experimental conditions and the inadequateness of decoherence, in most cases we cannot determine whether a history family precisely meets the consistency or decoherence conditions. Hartle believes that as long as the decoherence functional of this history family can approach 0 with an experimental undetectable approximation degree, it can be regarded as approximately consistent (approximately consistent), which does not hinder the original theoretical analysis; and

because it is extremely close to 0, within its sufficiently small neighborhood, there must be a point that is precisely equal to 0. Critics argue that, on the one hand, the claim that a point precisely equal to 0 can be found within a sufficiently small neighborhood has not been strictly proved from mathematics; on the other hand, no matter how close the approximation is, it still violates the basic axioms of probability, causing the decoherence functional $D(\alpha,\alpha') = Tr\left(\hat{C}_{\alpha}\rho\hat{C}^{*}_{\alpha'}\right)$ to not be strictly regarded as "probability", thus becoming meaningless.

3 The mode of specific selection and decomposition of the projector can be changed, as long as the relations in the above table is met and the consistency of the history family formed is guaranteed.

5 The nomological interpretation of quantum wave function

Laws of nature are a key concept in the philosophy of science, but there are significant differences among philosophers of science regarding the necessity and meaning of this concept. Some people believe that laws of nature are indispensable and that they can replace the concept of causality; others think that the concept of laws of nature is not necessary at all in science and can be completely replaced by mechanisms, models, symmetries, or some other concepts. The current understanding of laws of nature can be broadly divided into two categories – the regularity approach and the necessity approach. If further subdivided, there are several versions such as strong realism (the law of nature is equivalent to the necessary connection between universals), mild realism (the law of nature comes from the dispositionalist understanding of properties), mild anti-realism (a law of nature is an axiom of the best explanation system), and strong anti-realism (laws of nature are the subjective generalizations of rules by the subject). In addition, we need to consider how to understand the laws of nature in specific sciences and the relationship between the laws of nature and mechanisms, causality, probability, counterfactuals, dispositions, models, and scientific explanations. In biology, ecology, and quantitative social sciences, the disputes over the above issues are even serious.

As mentioned earlier, how to understand the status of quantum states is a core issue in the philosophy of quantum mechanics. Due to the significant differences among various interpretations of quantum mechanics, these interpretations' different understandings of quantum states and wave functions will inevitably affect the analysis of related issues, including laws of nature. Which part of the quantum mechanics theory is the ontology? After determining the ontology, which part belongs to the laws of nature or has the feature of nomology? Is the law of nature a real and necessary connection between quantum ontologies, or is it just an axiom assumption in the best explanation system? What characteristics do the laws of quantum mechanics exhibit, and which of the current approaches to understanding the laws of nature do they resemble more?

The following analysis focuses on the nomological interpretation of the wave function of quantum mechanics proposed by some people of the Bohmian school. To defend this interpretation, supporters have put forward various arguments and carefully distinguished different meanings of the concept of nomology. This is not merely related to the several major approaches to understanding laws of nature that

DOI: 10.4324/9781003662860-5

already exist in the philosophy of science but also shows a different novel perspective. In particular, these new arguments could serve as a reference for re-examining the formalism of the quantum histories interpretation.

5.1 Primitive ontology and the nomological interpretation of wave function

The theoretical status of the wave function in quantum mechanics has always been a difficult issue that has failed to reach a consensus till now. In the past 20 years, some scholars of the Bohmian school have put forward a view that the wave function can be given a "nomological interpretation",[1] that is to say, the wave function "is nomology, but not ontology". Later, some people combined the relevant discussions on "primitive ontology" and extended it to the spontaneous collapse theory and the many-worlds theory (Allori et al., 2011; Allori et al., 2014; Chen, 2017, 2021); The discussion on the nomological interpretation of the wave function from the perspective of Bohmian mechanics can be found in many articles (Dürr, Goldstein and Zanghi, 1996; Goldstein and Teufel, 2001; Goldstein and Zanghi, 2013; Dürr et al., 2009; Allori et al., 2008; Allori, 2015, 2017, 2019, 2020).

To illustrate the particularity of the wave function, it is necessary to briefly review how to define the primitive ontology in a physical theory, especially how to understand the primitive ontology in quantum mechanics (Allori, 2013). One of the original intentions of proposing the concept of primitive ontology is to extend the classical approach of understanding physics to explain the world as much as possible, and thereby reflect the continuity between quantum mechanics and classical physics. According to this line of thought, one first needs to determine the primitive variable used to represent "matter" in the theory – the primitive ontology must be located in three-dimensional space (or in four-dimensional spacetime), be microscopic and local, and be the material that constitutes everything else. No fundamental physical theory's explanatory schema can do without the primitive ontology. Following the specification of the primitive ontology, additional ontological elements represented by non-primitive variables are required. Some ontologies represented by non-primitive variables are also needed. The role of these ontologies in the theory is to "manipulate" the time evolution of the primitive ontology in a dynamic way. For example, in classical mechanics, the position of a particle represents the primitive ontology, while the momentum of the particle is a non-primitive ontology. In classical electromagnetic field theory, the position of a charged particle represents the primitive ontology, while the electromagnetic field $\left(E, B\right)$ can be regarded as a non-primitive ontology or another primitive ontology coexisting with the charged particle; if it is the latter understanding, then the electric potential V and magnetic potential A are equivalent to non-primitive ontologies, controlling $\left(E, B\right)$.

From this perspective, the primitive ontology in quantum mechanics should also be similar, with matter represented by local variables in three-dimensional space or four-dimensional spacetime, and their evolution laws determined by the wave function. Of course, apart from approaches like Bohmian mechanics that take particles as the primitive ontology, one can also investigate this in other interpretations of quantum mechanics, such as some versions of spontaneous collapse theory and many-worlds theory. Currently, there are the following proposals (Allori et al., 2008, 2011; Allori, 2020: 218–219):

1 Particles: in the spontaneous collapse theory, particles can be regarded as the primitive ontology, and the wave function of standard quantum mechanics is modified to the stochastic equation of the Ghirardi–Rimini–Weber (GRW) theory, and then combined with the guiding equation of Bohmian mechanics to solve the motion of particles.
2 Mass density field (as a scalar field in three-dimensional space): in the GRW theory, the mass density is regarded as the primitive ontology, and its evolution law is given by the randomly evolving wave function; in the many-worlds theory, the mass density field is regarded as the primitive ontology, and its evolution follows the wave function that satisfies the Schrödinger equation.
3 Flashes (referring to the wave function collapse events in spacetime, with each event corresponding to a spacetime point): in the GRW theory, each spontaneous collapse event of the wave function in spacetime is regarded as the primitive ontology, and the position of spontaneous localization is the location where the "flash" occurs, with its evolution law given by the wave function of the GRW equation.

Scholars put forward the above statements not to determine once and for all which one is the only definite primitive ontology in quantum mechanics, but mainly to better grasp what kind of picture of quantum reality different approaches provide and to more clearly see their similarities and differences.

The following analysis is about how the supporters of Bohmian mechanics establish the nomological interpretation of the wave function under the premise of taking particles as the primitive ontology. Dürr, Goldstein, and others initially formed the idea of nomological interpretation of the wave function in the 1990s. The monograph *Quantum Mechanics Without Quantum Philosophy* published in 2013 systematically argued why the wave function has to be interpreted nomologically (Dürr et al., 2013). It is well known that the two main equations of Bohmian mechanics are the Schrödinger equation

$$i\hbar\frac{\partial\psi}{\partial t}=\hat{H}\psi$$

and the guiding equation

$$\frac{d\boldsymbol{Q}_k}{dt}=\frac{\hbar}{m_k}\mathrm{Im}\frac{\psi^*\nabla_k\psi}{\psi^*\psi}(\boldsymbol{Q}_1,\ldots,\boldsymbol{Q}_n,t) \tag{5.1}$$

When scrutinizing the status of the wave function in Bohmian mechanics, it is necessary to clarify two facts: on the one hand, the wave function serves as the solution to the Schrödinger equation, and on the other hand, it participates in the formation of the guiding equation. What is the relationship between these aspects? Is the wave function subjective, epistemic, or objective? What kind of physical process is the collapse of the wave function? Dürr et al. believe that these problems can all be solved through the nomological interpretation of the wave function.

First, consider the case where the wave function is "epistemic". Suppose a thing X can describe physical quantities or the results of some physical measurements. Then, the wave function is called epistemic if it is essentially equivalent to a probability distribution in the space formed by the possible values of X. The orthodox view clearly does not understand quantum mechanics in this way, because here X seems to be a hidden variable, and the orthodox view rejects any assumption of hidden variables, only acknowledging the wave function. That is to say, in the orthodox view, the wave function should not merely be epistemic. Similarly, in Bohmian mechanics, the wave function is not merely epistemic: the wave function clearly plays a role in deriving probabilities, and the square of its modulus gives the probability distribution of the configuration of the Bohmian system, but this is not the only role the wave function plays in Bohmian mechanics. A large amount of literature on the interpretation of quantum mechanics can basically be classified into the following three attitudes toward the wave function: (1) the wave function is everything (all physical reality), a typical representative being Everett's view; (2) the wave function is part of reality, and all physical reality is composed of the wave function and some other things; (3) the wave function does not represent any real physical entity, and we can eliminate it from the theory, thus avoiding the confusion about its strange properties, as some versions of anti-realism hold.

There are two important concepts in Bohmian mechanics: the wave function ψ of the entire "Bohmian universe", and the wave functions of its subsystems. In the Bohmian universe, all particles are controlled by this Bohmian mechanics equation, and it is precisely this global equation that determines their motion. However, in practical applications of quantum mechanics, people often only consider the wave functions of certain subsystems of this universe, rather than the global universe wave function ψ, after all, ψ is obviously extremely complex and difficult to handle. But fundamentally, the only truly "Bohmian" system in the Bohmian universe is the universe itself. We cannot arbitrarily infer that all its subsystems are Bohmian from the Bohmian nature of the entire universe, although the motion of particles within them is still controlled by the wave function in the manner of Bohmian mechanics. Here, the behavior of a part of a large system is determined by the behavior of the entire system. For the entire system, only the universe wave function ψ and its configuration $\boldsymbol{Q}$ are the data we have and all objective physical reality. The wave functions of its subsystems can be defined through these data.

From the above analysis, Dürr et al. propose the following view on the status of the wave function: it is universally nomological and nomic[2]; it is more like a law of nature rather than physical reality.

The influence of the wave function on the configuration trajectory of particles is reflected through the guiding equation (5.1). In a compact form, the guiding equation can be rewritten as

$$\frac{d\boldsymbol{Q}}{dt} = v^{\psi}\left(\boldsymbol{Q}\right) \tag{5.2}$$

Note that in Bohmian mechanics, there is no reverse effect of particle trajectories on the wave function – that is, the configuration does not affect the wave function in any way, and the evolution of the wave function itself is only given by the Schrödinger equation, but the Schrödinger equation does not directly contain terms with the configuration $\boldsymbol{Q}$. Although the Schrödinger equation is a core component of standard quantum mechanics, it does not directly contain information about the "position" or "configuration" of particles in reality.

Furthermore, Dürr et al. emphasize that the wave function $\psi(q)=\psi(\boldsymbol{q}_i,\ldots,\boldsymbol{q}_n)$ of a multi-particle system should not be understood as a field with peculiar properties in the real physical space, as advocated by supporters of various field theory interpretations. Instead, it can only be regarded as a "field" in the configuration space – the set of all possible configurations of the system – because as long as the number of particles is greater than 1, the configuration space is not equivalent to the physical space. Both equations (5.1) and (5.2) are equations of motion and are "laws" concerning the movement of particles. The role of the wave function here is merely to form or provide this law on the right side of the equation.

To argue for the nomological nature of the wave function in Bohmian mechanics, it can also be compared with the "Hamiltonian" H in classical mechanics. The wave function of an N-particle system is located in a $3N$-dimensional configuration space, rather than in a three-dimensional physical space. Similarly, in classical mechanics, the Hamiltonian H of a system of N particles is located in a $6N$-dimensional phase space, not in a three-dimensional physical space. In such a $6N$-dimensional phase space, the Hamiltonian function $H = H(q,p) = H(\mathfrak{X})$ of the system can be given, and the equations of motion can be obtained by differentiating it. In people's understanding of classical mechanics, generally, it is not believed that the Hamiltonian function in this $6N$-dimensional phase space requires additional special interpretation or questioning of its reality because it is merely a suitable mathematical tool for deriving the equations of motion.

By analogy, the status of the wave function is actually similar to that of the Hamiltonian function in classical mechanics, but there are differences in mathematical treatment: formally, it is closer to H in the classical Hamiltonian equation, which should be $\log\psi$ rather than ψ. If a certain form of "derivative" is abbreviated as der, then equation (5.1) and the Hamiltonian mechanics equation

$$\frac{d\boldsymbol{q}_k}{dt} = \frac{\partial H}{\partial \boldsymbol{p}_k}$$

$$\frac{d\boldsymbol{p}_k}{dt} = -\frac{\partial H}{\partial \boldsymbol{q}_k}$$

can be rewritten, respectively, as

$$\frac{dQ}{dt} = \mathrm{der}(\log\psi)$$

$$\frac{d\mathfrak{X}}{dt} = \mathrm{der}H$$

The two are obviously similar in form. Moreover, adding a constant to H does not change the equations of motion of Hamiltonian mechanics, and the two sets of equations of motion are completely equivalent physically. Similarly, in Bohmian mechanics, multiplying ψ by a constant (that is, adding a constant to $\log\psi$) does not change the velocity of the configuration and the equation of motion calculated from the wave function. This invariance under the action of a constant factor can also be compared to the electromagnetic gauge field, which will be briefly analyzed later.

A possible counterargument to this view is that law of nature or physical laws should be constant and not change over time, and should not be objects with dynamic evolution, but the wave function of a subsystem of the universe generally changes over time. In addition, laws should not be controllable by humans, but when preparing and manipulating quantum states, people can completely change the wave function of the system.

Dürr and other's response to this problem is that the only wave function we really need to focus on and consider fundamental is the universal wave function ψ. Any wave function ψ of a subsystem can be defined from ψ. Therefore, if the nature of ψ is fully grasped, then the nature of all other objects defined from ψ is also fully grasped, thereby solving the problem of the nature of the wave function of a subsystem of the universe. From the perspective of the universe (or the "universal" level), if the following Wheeler–DeWitt equation is accepted to hold:

$$\mathcal{H}\psi = 0$$

Then the solution of this equation – the universal wave function ψ – can satisfy almost every requirement of universal laws of nature: time-independent, no dynamic change, and not controllable by humans. Although the Wheeler–DeWitt equation may have various solutions under different boundary and initial conditions, as long as any specific conditions are determined, the solution is unique. This also conforms to the properties that laws of nature should have.

How can a time-independent ψ be used to obtain a time-dependent Schrödinger equation that can describe the time-varying wave function? In fact, these time-dependent Schrödinger equations that describe subsystems are only derivative and phenomenological, rather than the most fundamental equations. Dürr believes that even this time-independent Wheeler–DeWitt equation may not be the most fundamental equation. Note that for the Bohmian mechanics equation of the universe (5.2), the vector field v^{Ψ} on the right side, as the derivative, is expressed in terms of ψ. There is also the view that the situation where ψ satisfies the Wheeler–DeWitt equation is accidental and has nothing to do with whether the fundamental equation of motion takes the form of (5.2). One response to this is: if ψ is also accidental, then this, from the opposite side, indicates that the Schrödinger equation, which is a derivative product of it, is even less fundamental.

Subsequently, Dürr et al. introduce the term "quasi-nomological": If the universal wave function ψ is admitted to be nomological, then what theoretical status does the wave function ψ of a subsystem of the universe have? Just as in the previous analysis, since ψ is defined by ψ and the configuration of the external environment of the subsystem, its properties also need to be analyzed by these two. Dürr believes that once the theoretical status of the universal wave function is clarified, the status of ψ is secondary and subordinate to the former. In this case, ψ can be called "quasi-nomological", meaning that although it is difficult to directly regard the wave function of a subsystem as a typical law of nature, ψ still partially carries some nomological nature and participates in the generation of the law. It is more like a "non-primitive" ontology (entity) that is associated with the behavior of physical reality (that is, the primitive ontology) rather than being physical reality itself. However, the laws that control the configuration of particles establish the connection between the behavior of the subsystem and its environment – this is precisely what is reflected by the expression $\psi(x) = \psi(x,Y)$. The parameter Y on the right side of the equation represents the environment in which the subsystem is located; that is, the components of the universal laws corresponding to the environmental configuration are relevant to this situation. Therefore, the ψ defined by ψ also reflects this connection with the environment. This is the meaning of what Dürr et al. call the "quasi-nomological" nature of ψ.

If the invariance under the action of constant factors is examined, then in addition to the analogy with the Hamiltonian function of classical mechanics mentioned earlier, it can also be compared with the gauge potential in classical electromagnetic theory (Allori, 2017: 184–185). Allori compared the similarities between the gauge potential in classical electrodynamics and the wave function in quantum mechanics, and believed that from these similarities, the nomological nature of the wave function can be inferred. If the wave function is a projected object, then it is similar to the gauge potential in electrodynamics; and the gauge potential is law-like in classical electrodynamics; if the above two points hold, then the wave function is also law-like and nomological.

The traditional view holds that the fields E and B in electrodynamics describe physical degrees of freedom, while the "potentials" (including the electric potential V and the magnetic potential A) do not directly describe physical degrees of freedom. The electromagnetic potential is a component of the equations that "govern" the evolution of the electromagnetic field:

$$E = -\nabla V - \frac{\partial A}{\partial t}, \; B = \nabla \times A$$

Among the various transformations of the electromagnetic potential, the gauge transformation

$$V \to V + \frac{\partial \lambda}{t}, \; A \to A - \nabla t$$

does not affect the observable electric and magnetic fields. For this reason, electrodynamics is a gauge-invariant theory, and the electromagnetic potential is thus called the gauge potential. Once a specific gauge is chosen, the specific form of the gauge potential is determined. Some suitable choices can facilitate the solution in specific cases, but the electromagnetic field obtained does not change with the choice of gauge. This degree of freedom in choice means that each gauge potential is actually an equivalence class under gauge transformation.

Similarly, the wave function can be regarded as a ray in Hilbert space, and it is also an equivalence class under phase transformation. This situation is similar to the gauge transformation of the electromagnetic potential. Consider a wave function ψ and another wave function $c(t)\psi$ that differs from it by a time-dependent phase factor $c(t)$. Choose a "gauge" with respect to time

$$\left|\frac{dc(t)}{dt}\right| = 0$$

Then the behavior of ψ and $c(t)\psi$ under the Schrödinger equation will be exactly the same. In standard quantum mechanics, the wave function provides all the information about the time evolution of a physical system. The analogy with classical electrodynamics mentioned above indicates that the things defined in the sense of gauge are not the objects representing matter, so the wave function as a ray in Hilbert space should not be regarded as the best candidate for representing physical objects. And if the gauge potential can be regarded as nomological, then the wave function is also nomological. The significance of the wave function in quantum mechanics does not lie in its direct representation of matter, but rather as a component for generating the motion-evolution laws of other things. In standard quantum mechanics, the so-called "things" refer to the probability distribution of the physical properties of the object. The earlier discussion by Allori et al. holds that the probability distribution is more "primitive" than the wave function and should be the mathematical object representing matter (see Allori et al., 2008).

5.2 Strong nomological and weak nomological interpretations of wave function

After analyzing the arguments of Bohmian mechanics supporters, Eddy Keming Chen introduced two concepts of nomology: strong nomology and weak nomology (Chen, 2019).

Bohmian mechanics supporters draw an analogy between the wave function and the Hamiltonian function in classical mechanics, and the nomology they argue for belongs to strong nomology. In classical mechanics, the Hamiltonian function H in the Hamiltonian equation represents the kinetic energy of the system and the potential energy of the internal interaction of the system, and the Hamiltonian equation directly guides the movement of particles in physical space, which can be regarded as the "generator" of motion. From the perspective of the usual philosophy of science, theoretical elements like the Hamiltonian equation should belong to the law of nature. However, the Hamiltonian function is only a function in phase space and does not represent anything in physical space, so it is not a material entity. Therefore, in this sense, the Hamiltonian function is "strong nomology". In Bohmian mechanics, the guidance equation controls the real movement of particles in physical space, and the theoretical status of the wave function is only as the generator of the guidance equation, and the guidance equation has no inverse effect on the wave function; moreover, since the wave function is a function of the variables in the configuration space (rather than the variables in physical space) determined in advance by the theory about the system, it explicitly appears on the right side of the guidance equation, very similar to the Hamiltonian function in classical mechanics, so it can be considered that the wave function has the status of strong nomology in the Bohmian mechanics system. This understanding of the wave function can be called the strong nomological interpretation.

Similarly, if we consider that the law itself should not change with time, then we only need to regard the time-independent universal wave function that satisfies the Wheeler–DeWitt equation as the wave function with strong nomology in the system. Although it is time-independent, this does not mean that particles in the universe cannot move, because the time-independent wave function can still generate a non-trivial velocity field in the configuration space.[3]

Some scholars disagree with the strong nomological interpretation of the universal wave function by the Bohmian school. For example, Hicks believes that we require the "lawbook" to be supported by observation and experiment, and all the scientific experiments we conduct are carried out and observed in an isolated subsystem of the universe, which means that the law must be applicable to the entire universe and its isolated subsystems at the same time (Hicks, 2018: 996).

Critics argue that if the universal wave function is regarded as a law describing the movement of particles in the universe (in conjunction with the guidance equation), then the fact that it cannot be regarded as the movement law of the subsystem alone indicates that it is not a law of nature (Chen, 2017: 8–9). Another supporting evidence is that according to Łos–Tarski's theorem in the model theory of mathematical logic, among the statements satisfied by a structure of a first-order theory, only those containing exactly one universal quantifier can remain satisfied in the substructures of that structure. This means that from the perspective of model theory, such singular propositions without universal quantifiers cannot maintain satisfiability in substructures and thus are not suitable to be regarded as laws (Dewar, 2021). Are singular propositions or existential propositions merely necessary true propositions in the theory or laws with some kind of abstract necessity? This divergence involves different understandings of the essence of the law of nature.

In addition to the strong nomological interpretation, the more mainstream view in the literature interprets the wave function in terms of weaker versions of nomological interpretation. The weak versions do not require the wave function to assume the position of the Hamiltonian function in the law system of classical mechanics, nor do they require it to be simple or independent of time. The requirement of "having nomology" here can have a weaker standard. In the Humeanism version, laws are merely axioms in the best explanation system for the spatiotemporal mosaic. In some Humeanism philosophical frameworks, even so-called "deterministic chances" (Loewer, 2001) and some intrinsic properties (such as mass, charge) can be regarded merely as components of the best explanation system, without having a fundamental ontological status (Hall, 2015).

In the Humeanism version of the nomological interpretation, the most fundamental reality only consists of matter (particles or fields) in four-dimensional spacetime. The wave function is merely a dynamic variable that concisely and effectively summarizes the information on the spatiotemporal mosaic. Therefore, although the wave function is a component of the best explanation system, it is not necessarily timeless or independent of time. Even if the specific form of an individual wave function appears quite complex, as long as the entire explanation system containing it remains the simplest among all possibilities, it is acceptable.

Some people question that if only the complete information of particle trajectories is required, there might exist another theoretical system that can provide this information with much less information than the wave function, because the trajectory of a particle is merely a curve in the configuration space, while the wave function directly assigns values to all points in the configuration space. Moreover, the wave function does not follow the particle's trajectory, and the particle's trajectory cannot determine the specific form of the wave function (referring to the assignment of values in the configuration space). Supporters of the Bohmian school can counter this by arguing that what physicists directly obtain in real observational behaviors are only the position information of experimental tools and instrument pointers, and then use the wave function to explain and predict. Moreover, in most cases, physicists do not have disputes over the specific form of a system's wave function and will default to using the simplest expression in that situation. Thus, it can be seen that the comparison of various best explanation systems and the discussion of the problem of supervenience are actually quite complex and subtle.

One characteristic of the weak nomological interpretation is that it has fewer requirements for the universal wave function. In the strong nomological interpretation, the Bohmian school holds that the wave function of a subsystem is merely defined by the universal wave function and boundary conditions, initial conditions. As the defined term, its ontological status directly depends on the defining term. However, in the weak nomological interpretation, it does not need to be time-independent, simple in form, or determined in a simple way. It can be highly complex and time-dependent, as long as it is the "simplest" among all the choices. Of course, the "realist" degree of the weak nomological interpretation seems lower, but it can still be regarded as realist as long as the extended Humeanism framework can take a realist attitude toward laws and properties. Currently, there are only discussions of the weak nomological interpretation within the extended Humeanism framework. Whether it is possible to establish the weak nomological interpretation in a non-Humeanism framework remains to be studied.

In a certain sense, the weak nomological interpretation seems overly simplistic. Starting from such a weak restriction, can any theoretical entity with an uncertain status be regarded as part of the "law" system? Is this interpretation standard too easy to meet? If we apply this method to classical physics, we can not regard the electromagnetic field as a fundamental entity but merely as part of the law system. In this way, all the entities in classical field theory would only be particles, rather than "particles + fields". This seems to deviate from the common view – because in classical field theory, people often judge whether an interpretation is a realist interpretation by examining whether it can assert the ontological status of the field. (Chen, 2017: 9–10) Arguing from the "effective wave function of the subsystem" may still encounter difficulties in explaining quantum entanglement, such as the difficulty in explaining the entanglement between the subsystem and the external environment (Gao, 2017: 26–28). In response to this, supporters of Bohmian mechanics have also made counterarguments, suggesting that it is necessary to distinguish between "separability" and "strong separability", which Humeanism does not recognize (Bhogal and Perry, 2017). These scholars believe that the separability accepted by Humeanism only requires that any physical state be determined by the intrinsic properties and spatiotemporal relations of the spacetime points that constitute the entire Humean mosaic, without requiring that the physical state within a certain region be determined solely by the intrinsic properties of the spacetime points that constitute that region.

The last perspective worth mentioning comes from Halvorson (2019: 158). He proposed another weak version of the nomological understanding, suggesting that the role of the wave function in the theory could be that of a "proposition", or more precisely, a "rule" for generating true propositions. The theoretical role of a proposition is definitely different from that of names and variables, both of which are used to refer to existing things. No matter what stance is taken on the understanding of the law of nature, a law of nature can never be regarded as a thing and cannot be within the scope of a first-order quantifier in a physical theory. That is to say, according to the nomological view, the wave function is not a beable in the sense of John Bell. From this perspective, the wave function is weak nomology.

In summary, this chapter has discussed the primitive ontological approach of quantum mechanics, several different nomological understandings of the wave function and their analogical arguments, scholars' different views on the properties that a law of nature needs to satisfy, and some criticisms of this nomological interpretation. However, since we only aim to refer to the analytical perspectives here, we cannot thoroughly review all aspects of the related debates. To clearly present the relevant arguments, a brief review of these viewpoints is provided below (the table is compiled by the author):

1 Universal nomology:

   The component/parameter with inevitability and not evolving with time in the fundamental motion laws.

   Examples: The Hamiltonian function in the Hamiltonian equation of classical mechanics, the universal wave function satisfying the Wheeler–DeWitt equation.

   Remarks: After this part is given, the whole theory can be directly determined.

2 Quasi-nomology:

   It needs to be constructed from the universal nomology of the universal wave function. It serves as a component/parameter in the motion laws that are local, time-dependent, and artificially controllable for subsystems.

Examples: The wave function of the subsystem of the universe.

Remarks: It needs to be derived from the fundamental laws. After this part is given, it can partially determine the phenomenological theory of the subsystem.

3 Strong nomology:

It does not exist in the physical space and does not represent any physical entity but controls the evolution and motion of physical reality.

Examples: The Hamiltonian function in the Hamiltonian equation of classical mechanics, the universal wave function satisfying the Wheeler–DeWitt equation, the past hypothesis.

4a Weak nomology (axiom of a best explanation system):

The axioms in the best explanation system can be regarded as nomology.

Examples: It has a wide range of applications, such as the wave function in Bohmian mechanics, some versions of spontaneous collapse theories, and the many-worlds theory with a mass density ontology.

4b Weak nomology (mechanism of generating true propositions):

Within a theory, it is used for generating the true propositions and cannot be placed within the scope of quantifiers of the theoretical statements.

Examples: Same as above, it has a wide range of applications.

At present, the removal of the ontological status of the wave function from quantum mechanics has become a common choice among some interpretation schemes. Besides the supporters of Bohmian mechanics, other proponents of such a view include, for instance, Rovelli, the founder of the relational interpretation. Recently, when summarizing his long-advocated scheme, he explicitly expressed such an attitude: the relational interpretation itself is realist, rather than instrumentalist or subjectivist, but its attitude toward the wave function is anti-realist. He clearly states that the wave function is merely regarded as a computational device, used to calculate the probability distribution of some relative properties of a system in interaction (Rovelli, 2022: 158). This view is quite similar to the understanding of the wave function in the quantum histories interpretation.

Overall, the description of the fundamental material reality and its evolution laws in the world picture has become much more complex at the quantum physics level than in classical physics. These new views, which deviate from and intersect with the classical concept of the law of nature, reveal the still unclear and even somewhat obscure connotations of the quantum picture. Here, whether the understanding of the law of nature is strong realism, weak realism, or anti-realism is no longer a matter of different treatment of the ontological commitment and the representational relationship between theory and reality in quantum theory. To what extent does the arrangement and "labeling" of the theoretical role played by the wave function have any substantive significance for the interpretation of quantum mechanics, rather than being merely a matter of wording? It is not difficult to see that another significant aspect of this discussion lies in the fact that it forces people to seriously redefine what is meant by "nomology", "realism/anti-realism", and "theoretical representation".

Notes

1 The term "interpretation" used here is related to the definition given in Chapter 2, but there are also differences. The connection lies in that the work done by these scholars is indeed aimed at clarifying the ontological presuppositions of the Bohmian mechanics system and the relationship between the wave function and the ontology; the difference is that their arguments merely focus on analyzing the theoretical status of this one theoretical component, the wave function, and comparing it with the wave function realism view, and the purpose is not to provide a complete interpretation of the entire formalism. To be consistent with the original literature, this chapter also uniformly refers to the explanations of the Bohmian school scholars as "interpretations".

2 Some literature has used the term "nomic", which has the same etymology and related meaning as "nomological". "Nomic" approximately means "it is inevitable in accordance with the law", and "nomological" approximately amounts to "law-like".

3 Regarding the problem of the directionality/irreversibility of time, the "Past Hypothesis" might have to be added as a fundamental law of nature into the theory as a basic assumption.

6 The histories formalisms from the nomological perspective

To revisit and analyze the histories formalism of quantum mechanics, this chapter will refer to the basic idea of the nomological interpretation of the quantum wave function as described in Chapter 5 (also see Chen 2019; Dürr, Goldstein, and Zanghì, 2012). The quantum wave functions can be interpreted as nomological and non-ontological elements of Bohmian mechanics. This inspires us that, in quantum histories formalism, the multiple conditions/criteria of determining the possible families of histories, whose validity can be examined by reasonable tests regarding the properties of subsystems, can be argued as lawful or nomological, while the quantum histories can be argued as the fundamental physical reality.

We will specifically focus on this novel depiction of the formalism concerning quantum reality, which enables us to reconsider the numerous objections to the histories interpretation that were raised in previous chapters, and re-examine/resolve from the nomological perspective the several aforementioned ontological and logical issues of quantum histories interpretation. However, there still remain challenging issues in the understanding of quantum histories. In the appendix to this chapter, two examples of distinguishing consistence condition and decoherence condition are presented, which reveal the subtlety of the role that different selection conditions play.

6.1 Multiple selection conditions (criteria) in the histories formalism of quantum mechanics

The fact that most families of histories that meet the requirements for consistency or decoherence do not represent any real processes in the physical world and should therefore be viewed as unphysical is a significant criticism of the interpretation of quantum histories. On the other hand, these conditions also exclude some families of histories that appear to be able to represent some real physical processes. Finding an appropriate and reasonable "discriminating condition", "selection method", "probability assignment condition", or "selection criterion" (different use of terminology in the literature) is important, as it allows for the identification of the natural, physically meaningful families of histories from a vast array of possible

DOI: 10.4324/9781003662860-6

histories and the assignment of appropriate probabilities to the histories that fall into those families. Riedel et al. have commented on the consistency condition:

> ...we emphasize that decoherence is a dynamical physical process predicated on a distinction between system and environment, whereas consistency is a timeless property of a set of histories, a Hamiltonian, and an initial state.... In contrast, the mere requirement of consistency does not distinguish a preferred set of histories which describe classical behavior from any of the many sets with no physical interpretation.
>
> (Riedel et al., 2016)

Physicists have proposed some stronger or weaker selection conditions in addition to the consistency condition proposed by Griffiths and the decoherence condition proposed by Hartle and Gell-Mann. Deep insights into the re-understanding of quantum histories from the nomological perspective have been provided by the discussion of selection conditions or criteria for various senses of "consistency" in the histories interpretation in literature like Diósi (1994) and Halliwell (2009a), as well as more diverse selection conditions proposed in other related literature. The following is an examination of some representative selection conditions.

In the study of the interpretation of consistent histories, Halliwell incisively highlights a fundamental issue: how to ascertain the circumstances in which distinct probabilities can be meaningfully assigned to the various histories of a closed system and the predictions that these probabilities, which correspond to the various histories, can make under the appropriate conditions. Indeed, on the basis of the groundbreaking work of the pioneers of histories interpretation, physicists have put forth a variety of additional selection conditions/criteria, varied in strength (the definition of "strength" is given below) and producing unique mathematical and physical outcomes.

One of the first important steps in this direction was taken by Diósi (1994). He considered composite systems consisting of a number of subsystems that do not interact with each other. Ensuring statistical irrelevance between these subsystems is a fundamental criterion that should be satisfied by a fair selection condition. Diósi demonstrated that only a small subset of the potential probability assignment schemes could meet this criterion.

First, we recall Hartle and Gell-Mann's decoherence condition mentioned in Chapter 3

$$D(\alpha,\alpha') = Tr\left(\hat{C}_\alpha \rho \hat{C}^\dagger_{\alpha'}\right) = 0, \quad \text{for all } \alpha \neq \alpha'$$

and Griffiths' consistency condition

$$\mathrm{Re}D(\alpha,\alpha') = \mathrm{Re}Tr\left(\hat{C}_\alpha \rho \hat{C}^\dagger_{\alpha'}\right) = 0, \quad \text{for all } \alpha \neq \alpha'$$

The decoherence condition is clearly stronger than the consistency condition. Notice that if there are n histories in the family of histories, then the consistency condition has a total of $\frac{1}{2}n(n-1)$ sets of constraint equations and the decoherence condition has a total of $n(n-1)$ sets of constraint equations.

Nevertheless, there have been some weaker selection conditions. For example, the linear positivity condition has been proposed in Goldstein and Page (1995). First, the so-called quasi-probability is defined as

$$q(\alpha) = Tr\left(\hat{C}_{\alpha}\rho\right)$$

and then we notice that the relation between this and the probability

$$p(\alpha) = Tr\left(\hat{C}_{\alpha}\rho\hat{C}_{\alpha}^{\dagger}\right)$$

is as follows:

$$\begin{aligned} q(\alpha) &= Tr\left(\hat{C}_{\alpha}\rho\right) \\ &= Tr\left(\hat{C}_{\alpha}\rho\hat{C}_{\alpha}^{\dagger}\right) + Tr\left(\hat{C}_{\alpha}\rho\hat{\bar{C}}_{\alpha}^{\dagger}\right) \\ &= p(\alpha) + D(\alpha,\bar{\alpha}) \\ &= p(\alpha) + \sum_{\beta,\ \beta\neq\alpha} D(\alpha,\beta) \end{aligned}$$

When the decoherence condition is satisfied, the quasi-probability is equal to the probability. Thus, the linear positivity condition is defined as

$$\mathrm{Re}\, q(\alpha) \geq 0.$$

Another weaker condition is the partial decoherence condition (Halliwell, 2009a), which is defined as the absence of interference between each history within a family of histories and its negation:

$$Tr\left(\hat{C}_{\alpha}\rho(1-\hat{C}_{\alpha}^{\dagger})\right) = \sum_{\alpha',\ \text{only for those } \alpha'\neq\alpha} Tr\left(\hat{C}_{\alpha}\rho\hat{C}_{\alpha'}^{\dagger}\right) = 0.$$

The term "partial decoherence" refers to the possibility of interferences across histories even while each history is decoherent with its negation. As a result, the non-diagonal terms of the decoherence functionals acting on some histories might not be zero. The above equation is equivalent to the conjunction of these two requirements:

$$\mathrm{Re}\, q(\alpha) = p(\alpha),$$
$$\mathrm{Im}\, q(\alpha) = 0.$$

Nevertheless, we do not know the definite connection between it and the consistency condition; for a given family of histories, it may satisfy only one of them or neither. Specifically, the partial decoherence condition guarantees that the imaginary part of the function is zero, whereas the consistency condition does not guarantee that the real part of the decoherence functionals of all pairs of histories is zero (as the consistency condition satisfies).

It is not difficult to demonstrate that the partial decoherence condition is weaker than the decoherence condition but stronger than the linear positivity condition. Nevertheless, the relationship between the partial decoherence condition and the consistency condition is unobvious: the former cannot guarantee that the real part of the decoherence functional of all history branch pairs is zero (while the consistency condition can meet this) but can ensure that the imaginary part is zero (while the consistency condition cannot meet this), so we cannot imply any one of them from the other one – a given history family may only meet one of them, and it is also possible that the given family meet neither of them.

A concise summary of the relationship among the four conditions is as follows: the linear positivity condition is the weakest; the consistency condition and the partial decoherence condition are both stronger than it, and they have some intersection but are not equivalent; the decoherence condition is in the intersection of the consistency condition and the partial decoherence condition and is the strongest condition.

To show the four conditions in a unified and compact way, they can be reformulated as follows (Halliwell, 2009a: 484):

$$D(\alpha,\alpha') = 0, \quad \text{for all } \alpha \neq \alpha' \text{ (decoherence condition)}$$

$$\sum_{\alpha',\ \text{only for those } \alpha' \neq \alpha} D(\alpha,\alpha') = 0 \text{ (partial decoherence condition)}$$

$$\mathrm{Re}\,D(\alpha,\alpha') = 0, \quad \text{for all } \alpha \neq \alpha' \text{ (consistency condition)}$$

$$\mathrm{Re}\sum_{\alpha'} D(\alpha,\alpha') \geq 0 \text{ (linear positivity condition)}$$

Diósi and Halliwell establish that only the decoherence criterion, which involves composite systems, passes "Diósi's test" among the above requirements. We are compelled by this fact to reconsider their subtle differences.

Diósi's test: consider a composite system consisting of a number of statistically unrelated subsystems, then any probability-assigning condition that can be applied to the subsystems must be able to be applied to the composite system as well (Diósi, 1994).

Suppose there are two statistically unrelated systems A and B, with initial states $\rho^A \otimes \rho^B$, and the chain operator of the history is $\hat{C}_\alpha^A \otimes \hat{C}_\beta^A$, then the decoherence generalization of this composite system is:

$$D^{AB}(\alpha,\beta,\alpha',\beta') = D^A(\alpha,\alpha')D^B(\beta,\beta')$$

The decoherence condition passes the test while the consistency condition fails, because

$$\mathrm{Re}D^{AB}=\left(\mathrm{Re}D^{A}\right)\left(\mathrm{Re}D^{B}\right)-\left(\mathrm{Im}D^{A}\right)\left(\mathrm{Im}D^{B}\right)$$
$$\neq\left(\mathrm{Re}D^{A}\right)\left(\mathrm{Re}D^{B}\right)$$

As for the linear positivity condition

$$q^{AB}\left(\alpha,\beta\right)=q^{A}\left(\alpha\right)q^{B}\left(\beta\right)$$

Here, $\mathrm{Re}q^{AB}\left(\alpha,\beta\right)$ may not be greater than zero. Therefore, it fails the test.

Then consider the partial decoherence condition: given that both $p^{AB}\left(\alpha,\beta\right)$ and $q^{AB}\left(\alpha,\beta\right)$ are factorizable, so we have

$$p^{A}\left(\alpha\right)p^{B}\left(\beta\right)=q^{A}\left(\alpha\right)q^{B}\left(\beta\right)$$

Since $p\left(\alpha\right)=q\left(\alpha\right)$ holds for every subsystem, the above equation automatically holds for the composite system as well, so it passes the test.

In conclusion, we can exclude both conditions from the physical requirements of the statistical irrelevance of subsystems, but it is unable to make any distinction between partial and decoherence conditions in this case.

Halliwell then proposes a "reverse Diósi's test": consider a composite system consisting of a number of statistically irrelevant subsystems, then any probability-assigning condition that can be applied to the composite system must also be applicable to the subsystems.

Although the requirements of the reverse Diósi's test seem comparable to those of the original Diósi's test at first glance, they are fundamentally different in essence. First, one could argue that while the partial decoherence condition does not satisfy it, the decoherence condition clearly does. Consider the following equation:

$$p^{A}\left(\alpha\right)\sum_{\beta}p^{B}\left(\beta\right)=q^{A}\left(\alpha\right) \tag{6.1}$$

For any homogeneous history,[1] there is $\sum_{\beta}p^{B}\left(\beta\right)=1$ and thus $p^{A}\left(\alpha\right)\sum_{\beta}p^{B}\left(\beta\right)=p^{A}\left(\alpha\right)=q^{A}\left(\alpha\right)$, so it passes the test.

And if an inhomogeneous history occurs, we must also sum over α to get 1, i.e.,

$$\sum_{\alpha}p^{A}\left(\alpha\right)\sum_{\beta}p^{B}\left(\beta\right)=1 \tag{6.2}$$

Even though there are inhomogeneous histories in β, the test still passes if the histories of the individual α are homogeneous. This is because the first summation term on the left side of the preceding equation is 1, which also forces the second summation term to be 1. On the other hand, neither the first nor the second summation term may always be 1 if β likewise has an inhomogeneous history (just the product should be 1), and from equations (6.1) and (6.2) together, we get

$$q^A(\alpha) = \frac{p^A(\alpha)}{\sum_{\alpha'} p^A(\alpha')} \tag{6.3}$$

This is merely a condition demonstrating that p is compatible with q; nevertheless, it does not hold in general for individual subsystems and fails the test, since equation (6.2) does not ensure that the probability summation of the subsystems is equal to 1 for histories that are not homogeneous.

Halliwell points out that the failure of the partial decoherence condition to pass the reverse Diósi's test stems precisely from the particularity of inhomogeneous histories. The general rules that can be satisfied by the probabilities of those histories obtained by summing inhomogeneous histories in the absence of decoherence are still unclear. Is the case in which the summation of history probabilities occurs not equal to 1, i.e., $\sum p_i \neq 1$, universal, or does it occur only in special cases? It is unknown now.

In addition to ensuring that probability assignments are consistent between statistically irrelevant subsystems and the corresponding composite system, probabilities should also be assigned in a consistent and stable way under any perturbations that may arise in the real physical environment, both before and after the perturbations. The formalism of quantum histories should accurately and efficiently describe and deal with this type of real perturbation processes as well, since for open systems, quantum measurements themselves are nothing more than a kind of perturbation. This is an important point to note, even though the founders of quantum histories theory primarily dealt with the case of closed or isolated systems. Diósi (1994) raised this issue for the first time and showed that his "test of dynamical stability" was not satisfied by either the consistency condition or the linear positivity condition.

A simple case is enough to demonstrate this. A dynamical perturbation is exerted to the system near t_k at certain point in a series of time $(t_0, \ldots, t_n)$:

$$\delta H(t) = \delta\left(t - t_k - o^+\right) \sum_{\alpha^k} \lambda_{\alpha^k} P_{\alpha^k}(t_k),$$

Here, $t_k + o^+$ is the moment when the exertion of perturbation occurs, which lies at a very short moment after t_k; λ_{α^k} is a real number denoting a simple kind of external field. This equation then states that if the binary (i.e., it can only take either 0 or 1) projection operator $P_{\alpha^k}(t_k)$ has the value 1, then a sudden external potential field λ_{α^k} is exerted on the system. In this way, the history chain operator $\hat{C}_\alpha$ is perturbed near t_k and become

$$\hat{C}'_\alpha = e^{-i\lambda_{\alpha_k}} U_k^\dagger \hat{C}_\alpha,$$

where U_k is the unitary transformation factor:

$$U_k = U^\dagger(t_k) \exp\left(-i\sum_{\alpha_k} \lambda_{\alpha_k}(t_k) P_{\alpha_k}(t_k)\right) U(t_k)$$

$U(t)$ is the solution of the state evolution equation $\frac{dU}{dt} = -iHU$ when the Hamiltonian has not been perturbed. For the decoherence functional $D(\alpha,\alpha')$, it remains 0 after the perturbation:

$$D'(\alpha,\alpha') = e^{i\left(\lambda_{\alpha'_k} - \lambda_{\alpha_k}\right)} D(\alpha,\alpha') = 0,$$

so, the decoherence condition passes the dynamical stability test. The other three conditions are assessed again:

The consistency condition obviously fails the test because in the case that Re$D'(\alpha,\alpha') = 0$, but $D(\alpha,\alpha') \neq 0$, the real part Re$D(\alpha,\alpha')$ will not always remain 0 after the decoherence functional is multiplied by a complex factor.

The linear positivity condition also fails the test because despite the perturbation there are

$$\text{Re}q(\alpha) = \text{Re}Tr\left(\hat{C}_\alpha \rho\right) \geq 0$$

However, after the perturbation, whether $\text{Re}\left(e^{-i\lambda_{\alpha_k}}\right)Tr\left(U_k^\dagger \hat{C}_\alpha \rho\right)$ is positive or negative, closely depends on λ_{α_k}, which does not guarantee positivity.

The partial decoherence condition also fails the test, because in the two equations Re$q(\alpha) = p(\alpha)$ and Im$q(\alpha) = 0$,

$$p'(\alpha) = D'(\alpha,\alpha) = e^{i\left(\lambda_{\alpha_k} - \lambda_{\alpha_k}\right)} D(\alpha,\alpha) = p(\alpha),$$

i.e., while Re$q(\alpha)$ remains unchanged, $q(\alpha)$ may change as a whole, and there is no certainty that the imaginary part will be zero. The above-mentioned simple example has been the exclusive focus of case studies of dynamical stability test as of yet, and its physical implications need further exploration. However, one of the theoretical justifications of the histories interpretation's proponents is revealed by the fact that only the decoherence condition passes the test: the assignment of probabilities to the histories ought to be both consistent and unaffected by whether the system is measured or not.

It is possible to think of the imposition of an external field in the above-described form as a straightforward measurement or perturbation. This argues that appropriate determination conditions should reflect some form of robustness, which is different from the prior discussion of subsystems versus composite systems. The requirement that the chosen histories family remain classical is demonstrated by the Diósi test, the reverse Diósi test, and the dynamical stability test, but they are still inadequate to provide the kind of selection conditions that would, by themselves, select quasi-classical realms as we would expect. These three tests, along with a plethora of other potential variations, are conjectured to be merely particular instances of a certain type of test that reflects broader physical criteria.

The above discussion shows us the uniqueness of the decoherence condition – it is not only well geometrically characterized (the pure states correspond to the orthogonality of the set of states $\{\hat{C}_\alpha | \psi\rangle\}$) but is also dynamically stable (all the properties associated with orthogonality are preserved under perturbations of $\hat{C}'_\alpha \mapsto e^{-i\lambda_{\alpha_k}} U_k^\dagger \hat{C}_\alpha$).

There are likely two mechanisms at play here, given how crucial orthogonality is to determining the validity of the probability-assigning conditions (Halliwell, 2009a: 488–489). The first mechanism relates to conservation. When averaged over a wide range of space, the system-environment division and different kinds of local physical variables should both stay approximately conserved. The fundamental reason for the decoherence between individual histories is that there are states that remain invariant formulation under the action of the chain operator $\hat{C}_\alpha$, so that orthogonality between individual states of the form $C_\alpha | \psi\rangle$ is maintained (Halliwell, 2003). The second mechanism is of a statistical nature, as demonstrated in the case of huge particle numbers: in large systems, any typical pair of states has a tendency to be approximately orthogonal. The orthogonality of decoherence functionals has a special status since both mechanisms for producing decoherence are associated with the orthogonality of states.

There are still a good number of unanswered questions. As previously stated, the physical processes that render probability-assigning conditions physically compliant appear to favor decoherence over partial decoherence or consistency. This may suggest that, in the event that a physical mechanism exists that proves conservation or the presence of an external environment, then fewer requirements need to be met.

The question of whether there is an auxiliary condition that can provide the full physical decoherence mechanism (such as conservation and environmental influence) in combination with/as complement to the partial decoherence condition (or the consistency condition) can also be asked. This condition is not the same as the decoherence condition. In other words, it might be possible to "split" the decoherence condition into two parts: the partial decoherence condition (or consistency condition, etc.), and the other is some kind of auxiliary condition. However, the specific nature of the physical mechanism reflected in this auxiliary state, if any, remains unclear.

Apart from the previously examined conditions, physicists have also looked at the "global consistency condition", which is a stronger criterion than the decoherence and consistency conditions (Losada, 2014a, 2014b).

This condition is derived from the so-called contextual histories theory, which has been developed in recent years by several physicists from Latin America. At first glance, this formalism differs from the consistent histories formalism in that it treats events that can be transformed to the same moment by the unitary transform as a class of equivalence, rather than discussing sequences of history events and chain operators at different moments. These equivalence classes are then used to define "contexts" that satisfy the structure of distributive orthogonal complements. (Losada, 2014a, 2014b). They subsequently discovered and proved that the way to strengthen the decoherence condition in the consistent/decoherent histories formalism – that is, in obtaining the global consistency condition – is where the essential distinction between this theory and the consistent histories interpretation actually is. The condition is defined as follows (Losada, 2018):

$$\mathrm{Re}D(\alpha,\alpha')=0,\ \forall\alpha\neq\alpha',\forall\rho_0$$

It is a fairly strong condition that the consistency of the family of histories holds for any given initial state. It can be demonstrated that the criterion is stronger than both the consistency and decoherence conditions. Many desirable features characterize a family of histories satisfying this condition, including the following theorem: a family of histories satisfying global consistency cannot be deduced from two contradictory histories.

Let there be three histories α,β,γ, two globally consistent families of histories $\mathcal{F}_1,\mathcal{F}_2$, $\alpha,\gamma\in\mathcal{F}_1$, $\beta,\gamma\in\mathcal{F}_2$, and α,β are contradictory histories:

$$\alpha\wedge\beta=\beta\wedge\alpha=0$$

Then there exists no common initial state ρ_0 of $\mathcal{F}_1,\mathcal{F}_2$ such that both of the following equations:

$$P(\alpha\mid\gamma)=1 \text{ and } P(\beta\mid\gamma)=1$$

hold simultaneously.

By enabling predictions and retrodictions, this theorem suggests that the global histories condition effectively addresses the first and fourth ontology critiques we covered in Chapter 4 and the first two types of criticisms on logic. The global consistency condition has unavoidably strayed from Hartle and Gell-Mann's emphasis on the significance of the dependence on initial states, even though stronger conditions in and of themselves can lessen dependence on framework choice and the uncertainty in inference (especially when considering the universe as a whole in cosmology).[2] According to Hartle, how the quasi-classical world may be rendered from the quantum world and how quasi-classicality can be seen in spacetime, depends heavily on the universe's initial state. Thus, the influence of the initial state of the universe and the particular form of the Hamiltonian in the subsequent moments must be carefully analyzed in order to describe this evolving process by means of decoherent/consistent histories. Adopting the global consistency condition would explicitly reject this line of reasoning. Is there a selection condition that, while stronger than the decoherence requirement and weaker than the global consistency condition, nevertheless somewhat captures the significant impact of the initial state on the families of histories? This possibility has to be further investigated in the future.

Finally, we pay attention to some other conditions proposed by physicists for selecting history families and the physical significance behind them. The first one is the stability condition proposed by Clarke. The proposal of this condition mainly stems from the distinction between decoherence timescales and dynamical timescales. However, when applied to the quantum histories formalism, it mainly requires that the occurrence probability of a history should not change too much over time compared to the dynamical timescale, that is, it should remain relatively stable during the dynamical evolution process. According to Clarke's reasoning, history families that satisfy the stability condition may avoid the discontinuity in the ascription of properties mentioned in Chapter 4. Through the action of the decoherence mechanism, this stability condition can rule out the possibility of superposition of distinct macroscopic states. This condition requires that $t_s > \lambda t_d$, where t_s is the stability timescale defined by Clarke and t_d is the dynamical timescale of the system (Clarke 2001: 182–184).

The next one is the strong positivity condition (Boës and Navascues, 2017; Wilkes, 2019; Dowker and Wilkes, 2022). The so-called weak positivity requires $\forall\alpha D(\alpha,\alpha)\geq 0$ (compare with the positive definiteness condition: $\forall\alpha\forall\alpha' D(\alpha,\alpha')\geq 0$). And if decoherence functional, viewed as a matrix,[3] satisfies that it is essentially a positive semi-definite matrix, then it is known in the literature as strong positivity. This is apparently stronger than weak positivity. Conversely, if one starts with a decoherence functional on an algebra of events as the axiomatic basis for the physics of a quantum system, the condition of strong positivity, if adopted as one axiom, allows a Hilbert space to be constructed. This derived history Hilbert space can be shown to equal the standard Hilbert space in non-relativistic quantum mechanics and in finite unitary systems, in a physically meaningful sense (Dowker, Johnston and Sorkin, 2010). Boës and Navascues give an argument for strong positivity based on the composability of finite, noninteracting, uncorrelated systems (Boës and Navascues, 2017). Dowker and Wilkes define composition at the level of "quasi-systems", which differs from "systems" only in that a decoherence functional is not necessarily weakly positive (since the composition of two weakly positive systems may not be a system) and proves a series of theorems that show strong positivity is an appropriate condition for infinite systems and strongly positive systems is the unique set of quantum systems that is maximal and closed under composition, which weak positivity does not satisfy. The desirability of the requirement of composability lies in that physicians might consider any attempt to split the whole physical universe up into subsystems, because for now, it is necessary to make any progress from considering subsystems, both in isolation from and in interaction with others.

When Riedel categorized and outlined various selection conditions, he emphasized that those typically considered too weak hold merely historical and conceptual significance for understanding formalism and lack physical significance. The mapping of quantum states to outcome probabilities derived from history families satisfying only one of these weaker conditions exhibits excessive sensitivity to purely formal modifications (e.g., appending non-interacting auxiliary systems), rendering it unacceptable. Only history sets fulfilling sufficiently strong selection conditions can be assigned true probabilities. The pivotal distinction lies in the consistency condition. Conditions strictly stronger than consistency inherit its robustness.

Conversely, all conditions stronger than consistency are, to varying extents, devised to tackle the so-called "history set selection problem". Specifically, these conditions enable the assignment of true probabilities and aim to discern, within the extensive space of possible history sets, those aligning with our perceived external classical reality. These conditions may be classified into two primary categories: logic-based and those inspired by physical principles. For more logical and physical selection conditions, the reader may refer to the various examples listed in Riedel (2014); for thc considcrations from quantum Darwinism, redundant record and decoherence upon seeking selection condition, and the question of how to obtain objectivity when many observers access records of the past and agree on the sequence of events that constitutes the history, see Riedel, Zurek and Zwolak (2016). On the issue of how can two or more observers reach objective measurement results, see Li et al. (2018).

6.2 The ontological status of quantum histories and the selection conditions from the nomological perspective

It is evident from the analysis in the previous section together with the definitions of terms in Chapter 2 that each of the selection conditions of the families of histories actually corresponds to a certain formalism of quantum histories, and that all of these formalisms have not yet fully finished their own interpretation and cannot be regarded as complete theories in the sense defined previously[4] (except for the formalism of consistent histories, which currently has a "realm/framework-dependent ontology" interpretation by Griffiths that does not seem enough realist). Instead of referring to them as complete theories, it would be more appropriate for the moment to refer to them merely as "formalism of consistent histories", "formalism of decoherent histories", "formalism of linearly positive histories", and "formalism of global consistent histories", until they get adequately interpreted.[5] We must clearly identify the ontological elements and nomological relations associated with each of these formalisms and address a variety of questions raised previously.

Recall from the conclusions in Chapter 5 that, in the nomological interpretation of the wave function of Bohmian mechanics, there are at least four levels of views that regard wave functions as universally nomological, quasi-nomological, strongly nomological, and weakly nomological, respectively. No doubt that there are differences in the application range of these various interpretations. For instance, weak nomology can be adopted in a larger variety of circumstances, and quasi-nomology is limited to the subsystem wave function, whereas universal nomology is consistent with the understanding of the universal wave function.

In Bohmian mechanics, the wave function is seen as an element (which can also be thought of as a parameter) of the guiding equation that takes part in the construction of the laws/nomologies. We can distinguish between primitive ontology and the guiding equation that guides the motion of primitive ontology. In contrast to Bohmian mechanics, the case is somehow blurred in the formalisms of quantum histories. If we take each single history in the family of histories that satisfy the selection conditions as the most fundamental reality of the theoretical presuppositions,[6] can the various selection conditions be consequently considered as laws/nomologies? Furthermore, we should not forget to elucidate the theoretical status of the initial states and the Hamiltonians of the system.

Although this viewpoint has some credibility, it also faces lots of aporias. First, let's demonstrate its plausibility. It is evident that, once a history family selection condition or criterion has been fixed, it fulfills several interpretations of what constitutes "law/nomology".

To begin with, each selection condition can always be considered weakly nomological, as it is merely a part of the possibly best system to which it relates, if we adopt the view as in the best system version.[7] Suppose a law/nomology is understood as a constraint that limits the possible states of physical reality.[8] Then, in classical mechanics, the nomology amounts to demarcate some specific subsets in the phase space for each physical system. In Bohmian mechanics, the guiding equations delimit the possible spaces of trajectories of the system's particles in potential space; similarly, each type of selection condition is also sorting out/delimiting a series of families of histories describing the system that it recognizes as physically meaningful. These families of histories can then be assigned probabilities and be subject to propositional inference. If, as Halvorson alleges (Halvorson, 2019: 158), the laws can be understood as some kind of generators of true propositions about the system,[9] then each selection condition can obviously reflect this function as well, and this is right in line with one of the major original intentions of Griffiths and others in proposing the consistent histories interpretation – that is, to return to the "truth-preservation" of proposition reasoning in classical logic within the selected (consistent) framework. Similar to Halvorson's statements, if a law is viewed as a "theoretical device" (rather than a real experimental apparatus) for the input/output of legitimate data, then the wave function in the guiding equation only represents a chosen internal parameter necessary for the device's operation. In contrast, the selection condition of quantum history families can be utilized to assess the truthfulness of statements within a selected (consistent) framework and can be directly regarded as the theoretical device itself, although it also needs to input some external parameters – such as the initial state and the Hamiltonian information of the system (for the global consistency condition, even the initial state information does not need to be input). This can also determine the status of the two elements of the initial state and the Hamiltonian in the theory.

Furthermore, some difficult issues around the nomological interpretation of the wave function in Bohmian mechanics – such as the contingency, artificial controllability, and time-invariant evolution of the subsystem wave function, and the fact that the existential propositions about the universe wave function do not apply to the subsystems – will be naturally avoidable in the context of quantum histories. First of all, unlike the subsystem wave function, it is evident that the various selection conditions in the quantum histories formalisms are neither contingent, nor manipulable, and nor do they evolve over time. Second, the selection conditions are not any "laws" of necessity, unlike the universal wave function, which is essential to Bohmian mechanics. Rather, they are not some conceptually ad hoc notions that appear to serve just as true statements or specific initial states. Any constructed family of histories must satisfy one of these selection conditions so as to be considered a fundamental reality of the formalism they represent. These conditions are all universal propositions, though they are not always propositions of the first order (in the sense of predicate logic), and which order they should be taken to be propositions depends on which part of the ingredients of the interpretation is chosen as the primitive ontology. Any quantum history family that is constructed must be regarded as a fundamental object that can be discussed and processed by the theory, only after at least one certain selection condition is satisfied; otherwise, it is just a pure mathematical construction without meanings.

Next, we analyze some challenges that such a perspective may encounter. First, there is a possibility that the levels of reality of the theory are somehow vague. The properties of each history in a family of histories, such as "there is no quantum coherence between each history", and so on, are indeed reflected in the selection requirements if they are to be interpreted as laws or nomologies. However, this still does not clarify whether it is the various possible history families obtained by decomposing the Hilbert space at each moment of the system that represent the physical reality of the system, or it is each history within the history families directly represents the physical reality of the system.

Each history can be understood as a collection of physical properties or events that are assigned to certain subsets of four-dimensional spacetime points (a Humean mosaic). This collection of properties or events can then be altogether understood as a historical process. However, no single history within the family of histories can be directly constrained by the selection conditions. For instance, a history that is a member of one consistent history family might also be a member of another consistent history family that is incompatible with it, or it might even be a member of a history family that does not even meet the consistency condition (or any condition concerned). The selection conditions directly constrain which meaningful history families exist, and can only indirectly constrain the logical relationship between histories propositions and the probability of realizing histories in a history family, since the selection conditions can contain information of initial states that is externally given.

As a result, the theory appears to require dividing the underlying ontology into two levels that correspond to distinct laws/nomologies. Either the many-worlds interpretation or the analog to classical statistical mechanics may be used to explain each individual history in the family of histories that have been chosen as meeting the selection conditions (see the discussion in Section 3.2).[10] In fact, only if the unique quasi-classical framework from which deterministic classical laws can be derived is actually found, as Hartle et al. expect, and if the probability of realizing the maximumly quasi-classical history in this framework is close to 1, can then the confusion regarding the hierarchy of ontology mentioned above be avoided. However, as several critics have pointed out, we cannot even assert that the formalism of quantum histories contains only one suitable quasi-classical framework in this circumstance.

As the studies show, during the questioning of the nomological interpretation of the wave function in Bohmian mechanics, some scholars have raised doubt about whether there could be multiple levels of nomologies (Solé and Hoefer, 2019: 133–134). Solé and Hoefer mentioned that in classical mechanics and relativity, several well-known principles such as the Lorentz covariance principle, the conservation of energy and momentum principle, and the general covariance principle seem to be at a higher level of nomology than the dynamical equations. Their role is to restrict the possible forms in which lower-level nomologies can appear in the theory. Nomologies that do not satisfy these principles are considered unacceptable or awaiting reasonable reformulation. However, the problem lies in that these higher-level principles are not directly added as the axioms in the axiomatizations of classical mechanics or relativity, thus failing to bring more explanatory power to the best explanation system. Seemingly redundant, the existence of these principles is contrary to the simplicity required by the best explanation system in the Humeanism style. From the considerations about the hierarchy of nomologies, the relationship among the guiding wave equation, the Schrödinger equation, and the wave function in Bohmian mechanics cannot even form any unambiguous hierarchy, because wave functions, the parameters appearing in the guiding wave equation which is a nomology, are governed by another nomology, the Schrödinger equation. There are still some possible responses to this. For instance, if the universal wave function, which takes the particles' position and time coordinates as variables and can determine the particles' velocities, is regarded as the only nomology, then it would be the only law of motion. In this version of the interpretation of Bohmian mechanics, the Schrödinger equation need not be considered a nomology anymore. Besides, there are some arguments that understand the wave function in Bohmian mechanics from the perspective of dispositionalism, but these will not be discussed here (Solé and Hoefer, 2019: 135–137). In regard to the histories formalism, the relationship between selection conditions and history family appears less ambiguous than in Bohmian mechanics. Furthermore, if we maintain that the conditions can be treated as a nomology constraining the possibilities/potentialities of constructing an acceptable history family, then there will not be crossing or overlapping of the ontological hierarchy/rank of them.

The second challenge is whether different quantum histories formalisms need to be compared in terms of "optimality" and, if so, according to what standards, from the perspective of the system of best explanation. Remember the questions raised in Chapter 4 and the preceding section. Contrary to what some may believe, we cannot simply assume that formalisms that use consistency or decoherence requirements already reflect the most conclusive and definitive understanding of histories formalisms and do not require further development. In fact, none of the current selection conditions – the decoherence condition, which eliminates some real physical processes represented by inhomogeneous histories, or the consistency condition, which fails the dynamical stability and Diósi's tests – can truly account for all the histories that we typically consider to be real and physically meaningful. See the examples in the appendix to this chapter for a more thorough explanation of the subtle distinctions between the consistency and decoherence conditions. The reasons for this situation are intricate (see Halliwell's analysis at the end of the previous section). It suggests that we are far from a comprehensive understanding of the temporal correlation/entanglement of quantum events and a systematic analysis of the quantum histories of those open systems interacting with their environment (Schlosshauer, 2004: 1301–1302).

Furthermore, none of the aforementioned conditions are optimal because none of them are able to distinguish the distinct quasi-classical framework that our current reality may have. This means that further research and creative effort are needed to advance our understanding of the formalism of quantum histories, which is still an unfinished work. It's possible that the formalism of consistent or decoherent histories does not, as some of its proponents stated, perfectly address the majority of interpretive problems and reach the conclusion of the theory's evolution. As was claimed in Chapter 2, the interaction between the theoretical extensions and philosophical interpretations should lead to significant future advancements in many (current and future) quantum histories formalisms.

Moreover, if it is believed that there exists a unique quasi-classical framework in this world where we live, then all the above-mentioned conditions are still unable to pick it out, and thus none of them can be regarded as "optimal". The research on the formalisms of quantum history is still unfinished. Perhaps it is not as its founders believed that the consistent histories or decoherent histories formalism has solved most of the interpretational problems. Nor has it reached the end of its theoretical development. As mentioned in Chapter 2, a series of (existing and yet-to-come) quantum histories formalisms need to be substantially advanced in the mutual promotion of interpretation and theory extension.

A detailed discussion along the lines presented here will be provided for some recent generalizations of quantum histories in Chapter 8. The examples include the measure of histories (coevent as an anhomomorphic valuation mapping that is the characteristic function of the history) and the duality of histories (superdensity operators and pseudo-density matrix), which illustrate how we could handle the ontology as potentiality and the dynamical law as constraints to the possibilities/potentialities of organization of the fundamental reality or even the logical relations of the propositions of reality.

6.3 Ontological and logical issues of quantum histories re-examined from the nomological perspective

Chapter 4 reviews several ontological and logical challenges to the quantum histories interpretation. Based on the discussion in this chapter, we can provide some responses to these challenges.

Let's first look at the ontological challenges. The quantum histories formulation highly depends on the selection of the framework when describing the attributes of physical systems, physical events, and physical processes, and it is difficult to ensure the persistence of the quasi-classical domain in this theoretical expression. The determination of consistency and the probability calculation of the history family, as well as the dependency on the external input of the initial state and Hamiltonian information, further weaken the independence of the theoretical ontology and may lead to a more puzzling ontological hierarchy.

First of all, we admit that this is a very different ontological picture from that of classical physics. Firstly, the historicized ontology given by the quantum histories formulation has a dual degree of freedom. On the one hand, even under determined selection conditions, the construction of the history family still has a large degree of freedom, thus allowing for multi-angle and multi-faceted characterizations of physical processes and physical attributes. This is convenient and easy for dealing with some quantum paradoxes related to time and non-commutativity. On the other hand, we can freely choose different selection conditions from a higher level according to our needs, thereby making the allowed history families themselves possess different properties under different conditions. Under conditions that are stronger than or equal to consistency, classical probability can be assigned to history propositions and classical propositional logic can be applied therein. Under even stronger constraints, it can also be applied to more complex situations of composite systems, or the inferences about the future/past through history families can demonstrate more convergence.

Some people may think that the quantum histories formalism is both too weak and too restrictive, but this critique itself is derived from the worldview based on classical physics. We can instead argue that the quantum histories formalism has both flexible and rigorous aspects. When dealing with various problems involving time-ordered event sequences, the quantum histories formalism has the convenience that other formalisms are difficult to rival, such as the advantage of directly avoiding paradoxes induced by non-commutativity by just checking the consistency.

Similarly, the absence of observers provides the liberty that observations and measurements are no longer essential for describing quantum processes, which allows us to discuss whether certain macroscopic things exist without worrying about whether we have aimed an instrument at it for measurement, but the price to pay is that we can thus legitimately discuss a large number of quantum histories that never actually occur. However, we think that this price to pay demonstrates precisely the particular value of the quantum histories interpretation.

Next, look at the logical difficulties. Starting from the propositions of the history family that have been settled to meet certain conditions, it seems that we can hardly make any reliable judgments about the events, propositions, and processes in another history family (whether it is compatible with the previous history family or not) as usually desired. This uncertainty across history families is the main source of the logical difficulties. But as the idea of Sorkin's quantum measure theory (to be introduced in Chapter 8) suggests, logic itself should be empirical and even dynamical in some sense. Either empirical content is embedded into the rules of logical reasoning (the so-called "dynamic logic" contained in quantum measure theory), or the scope of logical inferences is constrained in advance based on external empirical data, and then classical logic is restored within this limited range. This is very different from the situation in classical physics, too. Here, it involves a profound connection between the historicized ontology, logic, and causality. We cannot elaborate thoroughly on this now. But it is worth mentioning that the long-debated issue of the empirical nature of logic since Quine, Putnam, and Finkelstein, is not a problem only appertaining to philosophy of logic. From the perspective of historicized ontology, it will become an internal problem within physics. A predicate (=event=description) itself is a physical object embedded with temporality.

Finally, from the perspective of nomology, the historicized ontology directly incorporates the sequence of time into itself, so nomology is no longer a dynamic element that controls the evolution of the ontology over time, but rather a constraint on the possibilities/potentialities of all evolving processes. Some processes are allowed, while others are excluded. But the most important thing is not to find that unique special case, but to reveal what can be possible in the physical world through different degrees and forms of constraint conditions, each capturing a certain scope of possibilities/potentialities. This is not unrestricted freedom, but rather like using lenses of different focal lengths to view distant or near scenery. It is not that we can see whatever we want, but we are not limited to viewing a fixed distance with a single degree of lens. This constraint breaks the limitation of individual moments, thus more prominently demonstrating the holistic characteristics shared by the historicized ontology and the nomologies as constraints on possibilities/potentialities.

Appendix to this chapter: two examples of distinguishing between the consistency and decoherence condition

This chapter introduces three "tests", exploring the differences between four selection conditions for probability-assigning. It was discovered that the other weaker versions failed all of the tests, with the exception of the decoherence requirement. It is interesting to note that, in some cases, the family of histories that the weaker conditions accept can represent sequences of events of actual physical significance; however, it is excluded by the decoherence condition. This does not imply that only the family of histories chosen by the decoherence condition is able to describe or represent the real physical processes. The generality of these scenarios is still unclear due to the constraints of the cases analyzed.

The first example concerns inhomogeneous histories (Halliwell, 2009a):

Consider a family of histories containing only two moments t_1, t_2, where the projection operators for the corresponding moments are $P_1 = P(t_1)$ and $P_2 = P(t_2)$. Construct an inhomogeneous history chain operator

$$C = P_2 P_1 + \bar{P}_2 \bar{P}_1$$

and its complement

$$\begin{aligned} \bar{C} &= \boldsymbol{I} - C \\ &= I_1 I_2 - P_2 P_1 + \bar{P}_2 \bar{P}_1 \\ &= \left(P_2 + \bar{P}_2\right)\left(P_1 + \bar{P}_1\right) - P_2 P_1 + \bar{P}_2 \bar{P}_1 \\ &= P_2 \bar{P}_1 + \bar{P}_2 P_1 \end{aligned}$$

This family contains two histories $\{C, \bar{C}\}$. It can be seen that these two chain operators have a natural physical meaning, i.e., if P_i means "the particle is located at $x > 0$ in one-dimensional space (e.g., set to the x-axis) at t_i", then C means "the particle is on the same side of the x-axis at both t_1, t_2", and $\bar{C}$ means "the particle is on the opposite side of the x-axis at both moments". The non-diagonal elements in the consistency criterion for this family of histories are calculated as follows:

$$\begin{aligned} \mathrm{Re}D\left(C,\bar{C}\right) + \mathrm{Re}D\left(\bar{C},C\right) = Tr\left(C\rho\bar{C}^{\dagger}\right) + Tr\left(\bar{C}\rho C^{\dagger}\right) &= Tr\left(\left(\bar{C}^{\dagger}C + C^{\dagger}\bar{C}\right)\rho\right) \\ &= Tr\left(\left(P_1 P_2 \bar{P}_1 + \bar{P}_1 \bar{P}_2 P_1 + \bar{P}_1 P_2 P_1 + P_1 \bar{P}_2 \bar{P}_1\right)\rho\right) \\ &= Tr\left(\left(P_1 (P_2 + \bar{P}_2) \bar{P}_1 + \bar{P}_1 (\bar{P}_2 + P_2) P_1\right)\rho\right) \\ &= Tr\left(\left(P_1 \bar{P}_1 + \bar{P}_1 P_1\right)\rho\right) \end{aligned}$$

Since $P_1 \bar{P}_1 = \bar{P}_1 P_1 = 0$, $\mathrm{Re}D = 0$, and thus the family of histories is consistent. This physical scenario only satisfies the consistency condition – not the decoherence condition – because it is evident that the imaginary part of the decoherence functional is not always zero. $\bar{C}$ is also an inhomogeneous history, but its physical meaning is clear, and it can be interpreted as "the particle undergoes a motion that strides over the origin of the x-axis". If the decoherence condition is required to be satisfied, then histories such as C and $\bar{C}$ are ruled out as "meaningless" and cannot be discussed.

The quasi-classical physical situation, in this case, is in fact related to "the problem of arrival time". The issue of arrival time has been explored in the literature as a significant example of how consistent histories formalism is used to describe quasi-classical physical processes (Halliwell and Zafiris, 1998; Halliwell and Yearsley, 2009; Yearsley, 2010). The problem concerns a particle in one-dimensional space (e.g., the x-axis), and if the particle is located at $x > 0$ and has negative momentum, ask: what is the probability $p(t_1, t_2)$ that it crosses the origin exactly in the time interval $\left[t_1, t_2\right]$? One quasi-classical treatment is to solve for the difference between the probability that the particle is located at $x > 0$ at t_1 and the probability that the particle is located at $x > 0$ at t_2. The particularity of this type of problem is that, theoretically, the probability in quantum mechanics should be able to be expressed as a trace $Tr\left(\hat{P}_\alpha \rho\right)$ for some positive definite value measure (POVM) operator $\hat{P}_\alpha$, whereas $p\left(t_1, t_2\right)$ cannot be expressed as a trace $Tr\left(\hat{C}\rho\right)$ for some operator $\hat{C}$ and the initial state. The eigenvalue of $\hat{C}$ may even be negative here. Decoherent histories (a somewhat comprehensive answer to such difficulties) were provided by Halliwell et al. through their use of complex potential energy expressions to approximate the calculation. A distinction between the requirements from various selection conditions on the imaginary part of the decoherence functionals is necessary for the complicated potential energy to exist.

The second example does not involve inhomogeneous histories (Priebe, 2012). In fact, finding a family of histories in which all non-zero decoherence functionals are fully imaginary requires merely matching the real and imaginary parts. Think about a spin system where there are state vectors at all three moments (the subscript numbers represent each of the three moments):

$$\left|v_1^1\right\rangle = \frac{1}{\sqrt{2}}\begin{pmatrix}1\\ i\end{pmatrix}, \left|v_1^2\right\rangle = \frac{1}{\sqrt{2}}\begin{pmatrix}1\\ -i\end{pmatrix}$$

$$\left|v_2^1\right\rangle = \frac{1}{\sqrt{2}}\begin{pmatrix}1\\ 0\end{pmatrix}, \left|v_2^2\right\rangle = \frac{1}{\sqrt{2}}\begin{pmatrix}0\\ 1\end{pmatrix}$$

$$\left|v_3^1\right\rangle = \frac{1}{\sqrt{2}}\begin{pmatrix}1\\ 1\end{pmatrix}, \left|v_3^2\right\rangle = \frac{1}{\sqrt{2}}\begin{pmatrix}1\\ -1\end{pmatrix}$$

Their density operators are, respectively,

$$P_1^1 = \left|v_1^1\right\rangle\left\langle v_1^1\right| = \frac{1}{2}\begin{pmatrix}1 & i\\ i & -1\end{pmatrix}, P_1^2 = \left|v_1^2\right\rangle\left\langle v_1^2\right| = \frac{1}{2}\begin{pmatrix}1 & -i\\ -i & -1\end{pmatrix}$$

$$P_2^1 = \left|v_2^1\right\rangle\left\langle v_2^1\right| = \begin{pmatrix}1 & 0\\ 0 & 0\end{pmatrix}, P_2^2 = \left|v_2^2\right\rangle\left\langle v_2^2\right| = \begin{pmatrix}0 & 0\\ 0 & 1\end{pmatrix}$$

$$P_3^1 = \left|v_3^1\right\rangle\left\langle v_3^1\right| = \frac{1}{2}\begin{pmatrix}1 & 1\\ 1 & 1\end{pmatrix}, P_3^2 = \left|v_3^2\right\rangle\left\langle v_3^2\right| = \frac{1}{2}\begin{pmatrix}1 & -1\\ -1 & 1\end{pmatrix}$$

Thus, the histories family with three moments is formed as

$$\{D\} = \left\{\begin{matrix}P_1^1 \\ P_1^2\end{matrix}; \begin{matrix}P_2^1 \\ P_2^2\end{matrix}; \begin{matrix}P_3^1 \\ P_3^2\end{matrix}\right\}.$$

The chain operator consisting of the corresponding density operators can be abbreviated as $C_{i,j,k} = P_1^i P_2^j P_3^k$, where the index i, j, k are taken to be either 1 or 2.

Take the initial state as a simple one

$$\rho = \begin{pmatrix} 1 & 0 \\ 0 & 1 \end{pmatrix},$$

then the decoherence functionals become

$$D\left(C_{i,j,k}, C_{i',j',k'}\right) = Tr\left(C_{i,j,k}\rho C^{\dagger}_{i',j',k'}\right) = Tr\left(C_{i,j,k}C^{\dagger}_{i',j',k'}\right).$$

The calculation shows that the only non-zero non-diagonal terms are those that satisfy $i = i', j \neq j', k = k'$, which are all purely imaginary numbers, for example,

$$D\left(C_{1,1,1}, C_{1,2,1}\right) = \frac{i}{4}$$

Hence, the family of histories is consistent, but not decoherent.

Notes

1 Briefly speaking, a (homogeneous) history is represented by a time-ordered string of projections; if a history is by sums of such strings, it is named as inhomogeneous histories.

2 The existing assessment of the initial states of quantum histories has pointed to a possible conflict between the consistency of cross-framework inferences on quantum histories propositions and the histories formulation of quasi-classicality on the cosmological scale.

3 The elements of the $|\mathfrak{B}| \times |\mathfrak{B}|$ square matrix $M_{\alpha\beta}$ is $D(\alpha, \beta)$, where $\alpha, \beta \in \mathfrak{B} \subseteq \mathfrak{U}$, and $\mathfrak{B}$ is a finite event algebra on $\mathfrak{U}$.

4 A Complete Theory = Formalism with Interpretation (roughly speaking).

5 In different versions of the dynamic theory of spontaneous collapse, the basic dynamic equations are different (not only the parameters to be determined, but also the structure of the equations may be different), so that these "formalisms" actually correspond to a number of different "theories". The usual terminology is used only for the sake of simplicity to refer to a similar class of theories.

6 Even if quantum histories can be considered as a fundamental reality, it has an important difference from the primitive ontology described in Chapter 5 – the primitive ontology is considered to be a real physical spacetime fixed-domain, microscopic, and material reality, which is subject to the manipulation of the "non-primitive ontology" and undergoes evolution, but a quantum history itself is already a sequence of events/properties in four-dimensional spacetime, and each history already contains information about the evolution of the system and even about temporal entanglement (e.g., so-called entangled histories, see Cotler and Wilczek, 2015; Cotler and Wilczek, 2016; Cotler and Wilczek, 2017; Cotler et al., 2017). Also, since we can always consider the coarse-graining of any family of histories, the histories here are not always on the micro-scale (e.g., those histories in quasi-classical frameworks).

7 "…[Griffiths] most probably would subscribe to the weaker interpretation of explanation: for an occurrence to be explained we should show that it may happen given the laws and the circumstances (it is consistent with the laws and the initial conditions)". (Bigaj, 2023)

8 One related idea is raised independently from another context, see Salimkhani and Rolffs (2022). They "propose an alternative notion of scientific progress that metaphysical

investigations can and do latch on to, namely progress understood as exploring and constraining theory space". However, this article is mainly aimed to argue against McKenzie (2021) on naturalized metaphysics. Also compare Ruyant (2023): "the world is populated by situations, each characterized by a coarse-grained event space (its context) and a modal structure of events. …Once a situation is identified, hypothesized or intentionally created, quantum theory can be used to know its modal structure. But the theory does not tell us what situations there are in the world. Only direct experience does".

9 "…if we were to regiment the nomological view, we might say that the wave function plays the theoretical role of a proposition, or perhaps of a rule for generating propositions. The theoretical role of propositions is, of course, quite different than the theoretical role of names or even variables, both of which are used to denote existing things. Thus, only by stretching the word 'ontological' beyond the breaking point could we say that the nomological view is ontological. …No matter what view of laws we take, a law isn't a thing, and is not in the domain of quantification of a physical theory. Thus, according to the nomological view, the wave function is not a beable".

10 There is one possible solution to this problem for the formalism of decoherent histories, which was proposed earlier in Dowker and Kent, 1996: we may only consider the family of histories that satisfy the maximum degree of fine-graining of the decoherence condition in a given time interval. Since these families of histories are maximumly fine-grained, then in principle only one of the various projective decompositions can be satisfied. Under this presupposition, Kent argues that it is possible to adopt a "many histories" attitude to understand the individual histories in this identified family of histories. This is a way of interpretation of it.

7 Historicized ontology, scientific metaphysics, and philosophy of history

The temporality and a certain degree of perspectivism of the quantum histories interpretation and its formalism seem to have similarities with relational ontology, process ontology, and Leibnizian monadology. This chapter specifically analyzes the relationship, similarities, and differences between these versions of ontologies and the historicized ontology. Then, some issues in the philosophy of history are discussed in the second section. In the philosophy of history, some scholars have borrowed concepts and the worldview of quantum mechanics to understand the ontology of history since the 20th century. However, due to an insufficient understanding of the specific controversies of different interpretations of quantum mechanics, the influence on the philosophy of history often has misunderstandings and biases. From the perspective of quantum histories interpretation and scientific metaphysics, some of these misunderstandings can be reasonably resolved.

7.1 Comparisons between historicized ontology, relation ontology, monadology, and process ontology

Scientific metaphysics is a subfield of philosophy and also an intersecting discipline between metaphysics and science. It investigates the philosophical issues of core concepts in science. These concepts include, but are not limited to, natural kinds, laws of nature/nomology, causality, dispositions, possibility/necessity, explanation, reduction/emergence, grounding/fundamentality, and spacetime, among others. Traditional metaphysics focuses on metaphysical issues in everyday phenomena, such as the identity of objects, their components, and the problem of existence. In contrast, scientific metaphysics concentrates mainly on the abstract and universal concepts in science and utilizes cutting-edge progressions, methodologies, and intellectual materials in science for its research. What can be deemed as philosophical metaphysics of science can complement the philosophy of science by studying the ontological and metaphysical foundations of scientific concepts, for instance, "What is causality" rather than "How to understand causality". These two types of questions are not sharply distinct, since answering the former is bound to involve responses to the latter, nevertheless the emphasis that the former concerns is more essential.

As for physics, especially quantum mechanics, a large number of subtle, puzzling, intricate, and unexpected experimental and theoretical cases herein will have

DOI: 10.4324/9781003662860-7

substantive impacts on the above-mentioned scientific metaphysics issues, either promoting the debate on certain topics, eliminating some pseudo-problems, or opening up unprecedented views and sights. The quantum histories interpretation, due to its historicized formalism, inherently provides an essentially historicized ontology and thus will have a conceivable influence on the understanding of causality, spacetime, law of nature, etc. Besides, due to its prominent focus on history itself, though at the micro level, it will inspire philosophical and metaphysical thinking into macroscopic historicity, which can no doubt be regarded as a humanistic discipline.

In this section, we briefly discuss about the relations and comparisons between historicized ontology, Rovelli's relation ontology, Whitehead's process ontology, and monadology from the perspective of scientific metaphysics.

Whitehead's view on physics is deeply intertwined with his metaphysical framework, as articulated in his seminal work *Process and Reality* (Whitehead, 1929[1985]). Whitehead argued that the old Newtonian framework was inadequate to explain the dynamic, relational nature of reality revealed by modern physics. Quantum theory suggests a discrete, event-based structure of reality. These insights aligned with his processual view of reality as a network of interconnected events ("actual occasions") rather than static substances. Quantum indeterminacy resonated with Whitehead's emphasis on novelty and creativity in the universe. He argued that reality is not fully determined but involves spontaneity and self-creation at the micro level.

Whitehead's metaphysical "actual occasions" (the fundamental units of reality) are akin to quantum events – they are instantaneous, processual, and relational. Each occasion "prehends" (feels and integrates) its predecessors, creating a dynamic web of becoming. Whitehead came up with an organic model where reality is a living, evolving process. This reflects the worldview of field theories and entanglement in quantum physics. By taking the dynamism, relationality, and indeterminacy of physics seriously, he constructed a metaphysics where reality is fundamentally temporal, interconnected, and creative. Whitehead's view demands a metaphysics of process to fully articulate its implications.

Rovelli's emphasis on the relational ontology and event ontology of quantum physics aligns perfectly with Whitehead's process philosophy. "Authors who are not directly inspired by Whitehead often come up with interpretations of quantum mechanics that are strikingly Whiteheadian" (Desmet and Irvine, 2022). Rovelli explicitly claims:

In the world described by quantum mechanics, there is no reality except in the relations between physical systems. It isn't things that enter into relations but, rather, relations that ground the notion of "thing". The world of quantum mechanics is not a world of objects: it is a world of events. Things are built by the happenings of elementary events (Rovelli, 2017: 115–116).

There is no longer space that contains the world, and no longer time during the course of which events occur. There are elementary processes … continuously interact[ing] with each other…. The illusion of being surrounded by continuous space and time is the product of a long-sighted vision of a dense swarming of elementary processes (Rovelli, 2017: 158).

Maybe seeming dramatic to those who are not familiar with them both, Rovelli's relational quantum mechanics and Whitehead's process ontology share some profound conceptual affinities, particularly in their favor of a reality constituted by processes, events, and relations. Both frameworks challenge classical notions of objects with intrinsic properties, instead proposing that reality is fundamentally relational, dynamic, and event-based.

The core tenets of Rovelli's relational quantum mechanics include (Rovelli, 1996; Rovelli, 2025): (1) Quantum states are not absolute but exist only in relation to an observer or another system. There is no "view from nowhere" – properties like spin or position are meaningful only in interactions. (2) Quantum phenomena are understood as discrete interactions (events). (3) Unlike Bohmian mechanics, relational quantum mechanics denies that unobserved particles have determinate properties. Reality is constituted by the relations between observed events. (4) Measurement outcomes depend on the entire experimental setup, aligning with quantum contextuality (e.g., the Kochen–Specker theorem). Both Rovelli and Whitehead see reality as a web of interactions rather than a collection of objects that can be isolated and separated. A particle is regarded as a node in a network of relational events or a pattern of events. Relational quantum mechanics denies that quantum systems have properties outside of interactions (no hidden variables) and accept the fact that different observers can have different but equally valid descriptions of a quantum system.

Rovelli's relational quantum mechanics and Whitehead's process ontology converge on the idea that reality is not made of "things" but of dynamic, relational events. This synergy suggests that Whitehead's metaphysics may find empirical support from relational quantum mechanics. The event-relational ontology can be developed as a contra-substantialist understanding of physics and reality itself.

Epperson pointed out that in Whitehead's philosophy,

> the universe is described as an integrated system of serial routes (histories) of actual occasions that is, a global history of coherently integrated local histories. This is precisely analogous to the concept of local and global system histories in the decoherent histories interpretations of quantum mechanics. Coherence at all scales derives from the fact that the internal relations of every actualization constitutive of each local history are logically conditioned, such that the orders of causal relation and logical implication are always correlated.
>
> (Epperson and Zafiris, 2013: 25–26)

In quantum histories interpretation, "history" implies an ongoing process not necessarily connoted in the term "state". In quantum mechanics, the evolution of "state" and "history" is characterized as "fluid potentiality punctuated by an ever-expanding lineage of discrete actualities/events/facts, each novel actuality being subsequent to and partially consequent of all actualities antecedent to it" (Epperson, 2004: 82). And the concept of a "history" implies an actual past and a presupposed, subsequent, and consequent actual future. The alternative states of the reduced density matrix of any measurement can be characterized as alternative potential histories of the universe (Epperson, 2004: 83).[1]

Robson further suggests that the mathematical framework for consistent histories can be incorporated into relational quantum mechanics, concerning the behavior of stable facts of the latter (Robson, 2024).

As for monadology, McDonnell proposes a possible synthesis of Leibniz's theory and the quantum histories interpretation, which is called "Quantum Monadology" (McDonnell, 2019). Leibniz's celebrated theory of monads conceives they are simple, mind-like substances with a complete individual concept encoding all properties. Physical reality arises from synchronized monadic perceptions, not causal interaction. Moreover, monads can evolve independently but in perfect coordination. Apparent causality is an emergent feature of monadic perceptions.

Monads have some evident similarities with quantum histories. As for logical structure, both use propositional frameworks. The monadic harmony corresponds to non-interfering histories. The possible actuals of quantum histories resemble Leibniz's possible worlds. Surely there are contrasts between the two, such as the determinism from Leibniz's god-driven selection vs. the randomness from quantum probabilities. Monads are assumed as mental substances while quantum histories are believed as unrelated to mental existences. Due to all these observations, McDonnell deems monads as consistent quantum histories: each monad is a quantum history projecting possible actuals. The source of potentiality in choosing quantum history families contains all superposed possibilities.

These assumptions have some philosophical implications. It provides an alternative metaphysical grounding for the probabilistic quantum histories formalism and naturally explains classical emergence via decoherence and coarse-graining procedures, thus reconciling the Leibnizian version of idealism with quantum mechanics. Nevertheless, the information gathering and utilization system (IGUS) problem of how to define boundaries of observing systems still remains, and the actualization of histories still relies on an external selector (akin to Leibniz's god).

Ruyant envisages a broader picture of the relations between several interpretations of quantum mechanics: "Although there is no space for a full analysis, it should be noted that there are similarities between the present approach and the relational interpretation, modal interpretations, including perspectival ones and QBism" (Ruyant, 2023: 47). But the program of research on a possible reconciliation can only be carried out in future work. For a Bayesian account of quantum histories formalism, also refer to Marlow (2006).

7.2 Historicized ontology and philosophy of history in perspective of scientific metaphysics

With the establishment of quantum mechanics in the first half of the 20th century, a series of novel ideas and concepts aimed at depicting the microscopic world flooded the intellectual sphere. The impact was not limited to physics and its neighboring natural science branches; in fact, many fundamental concepts in the humanities and social sciences were also indirectly affected by the intellectual shock brought about by the quantum revolution, such as history, sociology, political science, and economics. The basic properties of quantum-scale objects, such as probability, observer participation, the uncertainty principle, complementarity, contextuality, non-locality, and non-classical logic not only challenged the classical causal structure, the dichotomy of subjectivity and objectivity, determinism, and even the existence of objective laws in the traditional world picture but also opened up an unprecedented and astonishing epistemology horizon. The fundamental issues that philosophy of history – as an important branch of philosophy – is concerned with are also related to this. For instance, whether history has regularity and objectivity, whether it is deterministic, and how historical knowledge can be true. This section briefly analyzes the changes brought about by the quantum mechanics revolution to the philosophy of history, especially regarding the metaphysics of history, and by drawing on the quantum consistent/decoherent histories interpretation and Hartle's relevant discussions, provides some meaningful references to respond to the above questions.

Philosophy of history is a discipline that conducts bottom-level reflection on the basic concepts, methodologies, disciplinary foundations, and the nature of "history" itself. It has both the aspects of history and the concerns from philosophy. Its essential content includes: what the research object (historical ontology) of history is, whether history is a science with objectivity or a highly subjective and creative art, whether historical knowledge is reliable and effective, whether historical facts exist, whether laws exist in history, whether humans can effectively understand and grasp real history, the rationality of historicism, the similarities and differences between historicity and temporality, the narrative of history and the generation of meaning, the understanding and interpretation of history, and many other issues on. The emergence of these issues marks the self-reflection of history as a discipline when it has developed to a certain extent. The problems that philosophy of history needs to face and handle are, first and foremost, the fundamental issues of history, but they also cannot be separated from philosophical inquiry and analysis. The author believes that these issues can only be answered by drawing on the intellectual resources of the philosophical community and even the entire human culture.

Modern philosophy of history, which began in the Enlightenment era, was born along with the emergence of classical physics, especially classical mechanics. After Francis Bacon's discussion on the goals and methods of scientific research and his call for empirical and experimental methods, and after the formation of a Baconian scientific research paradigm in the form of Newtonian mechanics in the 17th century, various "humanities and social sciences" were stimulated. However, the philosophy of history produced by this Newtonian mechanical worldview and the spirit of "mechanism" will be severely challenged in the new century's physics revolution. The research results of quantum mechanics on the physical laws of the microscopic world indicate that Newtonian classical mechanics is not a universal truth applicable everywhere, but only approximately and locally applicable to the macroscopic, low-speed world of human life.

At the beginning of the 20th century, quantum mechanics, upon its establishment, sparked intense debates among physicists, philosophers, humanists, and social scientists. People were unable to grasp the meaning behind the numerous experimental phenomena verified and predicted by quantum mechanics that were surprisingly different from the classical physics picture, as well as the novel abstract concepts in this theoretical frame. Especially the wave-particle duality and intrinsically probabilistic behavior of microscopic particles, as well as the influence of the observer's observation upon the motion of microscopic particles, challenged radically the causal theory in traditional philosophy and the dichotomy of subject and object between humans and the external world. From the micro level and thus also at the macro level, is it no longer possible to expect objectivity, causality, and determinism? Then, does the subjective participation of the observer become an inevitable element in the process of human beings' comprehension of the entire world (which includes nature and human beings themselves, as well as the entire history of their development)? And does all the natural and historical knowledge people obtain thereby become relativized, probabilized, and subjectivized illusions?

The notorious "uncertainty relation" (also known as the uncertainty principle) of quantum mechanics also reflects a major predicament in historical research, that is, the significant impact of the researcher's subjective preferences and emotional projections on the research conclusion. Therefore, there has always been a long controversy in the philosophy of history about whether history is an art, a technique, or a science. This seems to echo "the act of measurement has an undeniable influence on the microscopic object being observed" in quantum mechanics. Especially, between two quantum-scale objects, the so-called act of observation itself means a significant interaction on them, that is, an inevitable mutual interference and influence.

In addition, there is another quantum mechanics effect that is "reflected" in historical research, that is, the coherence. The main difference between human activities and the material movement and evolution of nature lies in that the former is an activity with human subjective consciousness, while the latter does not have such subjective consciousness. History is the ultimate result and product of all human conscious and purposeful activities. Human behavior is full of variability, complexity, and even unpredictability, and these characteristics have an immeasurable impact on the course of history, just as Popper said, the Œdipus effect – human predictions may have a promoting or hindering influence on the actual course of history. Human subjective activities can affect the realization of historical laws and even have a coherence with the causal relations in history, thereby leading to uncertainty.

Although some historians and philosophers of history have touched upon and expounded on the similarities between the characteristics of quantum mechanics and historical research, these discussions are often based on the most well-known Copenhagen interpretation among various interpretations of quantum mechanics, which is popular in physics textbooks[2]. Actually, many physicists have been dissatisfied with the Copenhagen interpretation, especially because its fundamental starting point is the forced retention of the classical physical world, the Kantian agnosticism of the thing-in-itself, and an instrumentalist attitude toward physical theories. From Einstein, Schrödinger, de Broglie, and others, to later philosophers of physics who proposed their realist interpretations, all of them hoped to provide quantum mechanics a reasonable interpretation to eliminate such subjectivism, instrumentalism, and mysticism. However, on the one hand, their efforts still do not have a certain result, and it is difficult to reach a consensus – the dispute continues (see Chapter 1). On the other hand, because the Copenhagen interpretation was once dominant in the physics community, scholars outside physics are often influenced by it at first, obtaining a series of preconceived notions about quantum mechanics and quantum reality, believing that this is the final conclusion of the quantum mechanics interpretation recognized by the physics community, and ignoring the efforts of scholars holding other realist views and the interpretation schemes they have proposed. These achievements are not without enlightenment for us to examine and reflect on the postmodernism and subjectivism tendencies gradually fermented by the anti-realism ideas in the interpretation of quantum mechanics and their impact on the humanities and social sciences. The following will start from the quantum histories interpretation (decoherent histories interpretation) of quantum mechanics that has been analyzed in detail earlier to discuss Hartle's relevant discussions on the "comprehensibility" of the world and history (Hartle, 2016, 2021: 349–366). Through analysis, it is found that these discussions bring new insights to reflecting on the crisis of contemporary philosophy of history and its future way out.

The establishment of quantum mechanics brought about a revolution from the classical worldview to the quantum worldview, but the classical worldview still has a strong conceptual inertia, and many people still want to maintain and preserve it. The once widely accepted Copenhagen interpretation of quantum mechanics actually partially retains the basic concepts of classical physics; that is, there must be a so-called "classical" observer in the interpretation picture. Although the act of an observer observing a quantum object will affect the observed object, it is still a special existence that is "detached from the object", so it can be said that the Copenhagen interpretation essentially carries the remnants of the old physics worldview. This classical observer is similar to the historian in historical narrative who can stand outside the event and describe history from a "God's-eye view" (even if his narration and record may have an impact on the development of later history). Many physicists gradually realized the incompleteness and incoherence of the Copenhagen interpretation. The quantum histories interpretation, which belongs to the "neo-realism" and has been developed since the 1980s, is contrary to the Copenhagen interpretation. In the treatment of the observer, it adheres to the basic position of regarding the entire world as a quantum object to be depicted, refuses to retain the artificial division between the classical world and the quantum world (i.e., the "Heisenberg's cut"), and eliminates the existence of an observer outside the observed system, that is, it considers all observers and observed objects as components of the entire quantum system.

In the quantum histories interpretation, the most basic constituent elements of the world are not quantum states or individual quantum events, but "quantum histories", that is, a sequence of quantum events parameterized by time (a sequence of events). To describe a sequence of events or calculate the probability of their occurrence/realization, the "decoherence" condition must be met, that is, there is no coherence among the event sequences in all possible history divisions of events within a certain period of time; otherwise, we will not be able to describe them coherently. Of course, this is only for the behavior of particles at the quantum scale. The problem is, if the world as a whole obeys quantum mechanics, then why can we, as observers living at the macroscopic scale, still feel and discover the laws and rules of classical physics, and thus consider the world we live in as "comprehensible"?[3]

In the quantum histories interpretation, to explain this, it is necessary to introduce the pair of concepts of coarse-graining and fine-graining. Coarse-graining means to ignore some details in a sequence of events, such as ignoring some events that occur at certain moments or ignoring the specific details of certain spacetimes and objects. For example, when observing the motion of the moon around the Earth, if the minute movements of the internal components of the moon are not studied, then only the motion of its center of mass needs to be considered. During the coarse-graining description, many originally coherent and indescribable details lose coherence due to the mutual cancellation of phase factors during the continuous accumulation process, thus presenting the characteristics of classicality.

Here, it is necessary to recall the concept of the "IGUS" introduced by Gell-Mann and Hartle.[4] Whether it is humans, animals, and various living beings, or measuring instruments, a car, an intelligent robot, or even a certain material system in a given spatiotemporal region, they can all be regarded as IGUSes. As the name suggests, its main features are: (1) It can obtain information from the surrounding environment. (2) It can utilize the laws in the obtained information to update the model or schema it constructs for the surrounding environment. (3) They can act based on the predictions given by the schema and further acquire information from the process of action. For IGUS to exist and operate, three conditions are required: (1) There must be a certain degree of reliable nomology for them to construct schemas and complete various actions; (2) There must be a certain degree of randomness to enable their evolution; (3) There must be a certain degree of complexity to provide sufficient possible directions for their evolution. According to the current cosmological model, the quasi-classical world we humans currently live in, compared to the early and late universe, precisely meets these three conditions, and thus a large number of IGUS exist here. Therefore, we can answer the previous question: the reason why this world has (or can display) nomology lies in the fact that if this world were to operate without any rules, no IGUS could have emerged. Conversely, since various IGUS already exist in this world, it cannot be a completely chaotic collection of matter and spacetime without any rules. The reason why this world is comprehensible is precisely because we are quasi-classical objects within it, rather than transcendent objects outside and separate from it. Regarding the issue of predictability, it is precisely because the cognitive subject of physics – human beings – have adapted to the nomology of the specific universe they are in and survived in it that the physics we have developed is predictable (Hartle, 2021: 360). Similarly, because the cognitive subject of history – social groups in human civilization – have adapted to the objective reality conditions they are in and survived in it, the predictability of human history and the social-historical laws they have recognized is possible. The previous discussion seems to have an obvious "anthropic" flavor, but from a certain perspective, it indeed directly dissolves the problem of "why does the world have nomology and why is the world understandable", that is, the existence and cognitive ability of the questioner (as an information collection and utilization system) already indicates the answer. Regarding the philosophical reflection on history, the significance of such a conclusion lies in the fact that it shows that the uncertainty, subjectivism, and perspectivism positions in the debate on the laws and objectivity of history do not have direct support from quantum theory, and we do not need to fall into mystified problems based on some interpretation that relies on inexpressible things. On the other hand, the conceptual framework provided by the research on quantum mechanics interpretations (and the epistemological research extended from it), represented by Hartle's decoherent histories interpretation, already contains some basic elements related to this philosophy of history debates and thus can serve as a reference (rather than direct application) for the exploration of the philosophy of history issues. Quantum mechanics, as well as the quantum world we exist in, can and must maintain that the laws within it are objectively existent. The cognitive subject of history also follows this law.

Although in the decoherent histories interpretation, different quantum history frameworks may give different descriptive pictures, due to the decoherence of the quasi-classical pictures within the history framework, they will present a complementary relation. As Hartle claimed, the quasi-classical realms are not the only decoherent set of coarse-grained alternative histories predicted by H, $|\Psi$. Since there is no finer-grained realm, some other coarse-grained realms may be incompatible with our quasi-classical world. Incompatible realms provide complementary descriptions of the universe, just like the changing histories of a particle's position and momentum. Both are necessary for a complete description. Quantum mechanics does not favor one over the other (Hartle, 2021: 359). However, for each cognitive subject IGUS in a specific history framework, they will provide a definite description of the event sequence of the world they are in. Even if these descriptions are incompatible, they will still form complementary pictures and become part of the complete history.

In fact, historical exploration will not deviate from science due to its complexity and limitations and turn into subjective, arbitrary fictional stories. On the contrary, it is precisely because of these characteristics that it has a scientific nature. First, even if a simple reductionist nomology exists, it does not mean that all problems can be solved. The reason why we cannot find a simple and clear theory of everything that can explain and predict everything is probably because "If it is short enough to be discoverable, it is too short to predict everything" (Hartle, 2021: 72). Second, the complexity and limitations of theories are advantages rather than disadvantages. Hartle pointed out in the context of quantum cosmology:

> The reason why complex adaptive systems can succeed in evolution and individual behavior is that they can recognize and utilize the nomology exhibited by the universe. Scientific theories can predict what these nomologies are and explain their origins. Theories can be used to estimate the feasibility of these predictions in terms of computation or measurement. By comparing different

> theories derived from the same data, we can know the reliability of our predictions. Therefore, the existence of such limitations does not represent the failure of the scientific enterprise. Limitations are an inherent nature of the scientific enterprise, and how to make divisions of limitations is an important issue in science.
>
> (Hartle, 2021: 108)

The humility of science toward the unknown is precisely the guarantee of its existing success and should never be regarded as a sign of failure.

Looking back at the history of human civilization over hundreds of thousands of years from the perspective of the Big History of evolution, human language is an important beginning mark in the history of human development, providing assistance for humans to enter a new stage of development from the material, social, and spiritual levels. The three most important intellectual awakenings in this evolutionary history are the three great leaps in human understanding of nature, the world, and themselves, namely the early "emergence of language" (enabling humans to cognize the world at the daily scale) and the modern Scientific Revolution (enabling humans to cognize the macroscopic and microscopic world), as well as the third cognitive revolution that began in the 1950s and mainly occurred in the fields of artificial intelligence and brain science (enabling humans to explore the depths of intelligence and consciousness). Following the cognitive revolution, the human-machine integration will reach unprecedented levels. With continuous advancements in biotechnology and artificial intelligence, humanity may transcend genetic constraints. The level of human control and utilization of the history of themselves and laws of nature through technological intervention may reach an unprecedented level. Does this imply the arrival of the transition from history in itself to history for itself? We cannot foretell yet. But since human lives as a special IGUS and an adapter in a complex environment, seeking, exploring, and utilizing the laws of nature and history to continuously develop and improve itself will be the eternal mission of humanity.

Notes

1 Another prospection worth mentioning here on the ontology of events, histories, and "propensitions" (a term from Maxwell N.) comes from Rudolf Haag's last writings (Haag, 2015: 233; also see Haag, 2019): "A full solution of this problem could open a wide perspective: the representation of possible histories as a category whose objects are events and whose directed arrows are propensitons describing causal links between events. It would include the description of possible space-times generated by the processes and dependent on the energy-momentum flow".

2 The confusion caused by the diversity of interpretations of quantum mechanics has sometimes misled the discussions in the philosophy of history (and other humanities), some linguistic factors, such as the technical jargons and abstract concepts of quantum physics, also pose difficulties for researchers from other fields. What is more troublesome is that since quantum mechanics itself is not the expected ultimate physical theory, it may continue to be challenged and developed. At that time,

> Given the trend in the development of fundamental theory, it is very possible that the disparity between human language and the language of fundamental physics will increase as quantum theory is replaced or extended. If that is the case, careful

> analyses of the relationship between human language and the language of physics of the kind sketched all too superficially in this essay will be increasingly important for clarity of understanding.
>
> (Hartle, 2007: 3116)

The situation of scientific terms and scientific rhetoric is more complex, involving the limitations of human language itself, the changes in concepts and terms brought about by theoretical development, and the paradoxes caused by the continued use of old terms. "Remember that human language can contain tacit assumptions that reflect the limited context in which it evolved" (Hartle, 2007: 3115). Only by using more precise physical language can the confusion caused by "linguistic excess baggage" be avoided. Borrowing physical terms in the humanities (spiritual disciplines) through analogies and metaphors may face double misplacement and deviation. This will not be elaborated here. See Hartle, (2007).

3 Einstein once lamented, "The most incomprehensible thing about the universe is that it is comprehensible". Hartle's title of this article echoes this famous quote.

4 It is worth noting that this concept can, to a certain extent, resemble the widely employed concept of "agent" in computer science, jurisprudence, and artificial intelligence. However, Gell-Mann and Hartle primarily focus on its physical ontological implications, without delving into issues such as activity capabilities or cognitive behaviors.

8 Conclusion and directions for future research

In this final chapter, we systematically examine and synthesize the key conclusions derived from the previous parts, while also proposing several important and interesting directions for future research that could extend our current topics. Specifically, we discuss (1) generalizations and variants of the historicized ontology, including the coevent formalism or quantum measure theory, the superdensity operators formalism, and the pseudo-density matrix formalism; (2) the physical extension of the application of quantum histories formalism, such as open quantum systems, infinite-dimensional cases, and classical/quantum relativity, and some interpretational and philosophical reflections. Finally, we give some remarks from the methodological considerations for future foundational research of quantum mechanics.

8.1 Possible future research directions – Part I: generalizations and variants of the historicized ontology

Based on the analysis and exposition in Chapter 6, recent generalizations of the notion of quantum histories, such as the measure of histories ("coevent", an anhomomorphic valuation mapping as the characteristic function of the history), the duality of histories (superdensity operators that can take as input any kind of operators but not just projectors), and other variations of the definitions around quantum histories formalism aimed at giving time and space a new treatment, will be discussed along the line.

The coevent formalism/quantum measure theory (these two names are adopted in different contexts by the theorists) regarded by its advocates as a realist interpretation, was introduced by Rafael Sorkin in Sorkin (2007), following his earlier work (Sorkin, 1994, 1997), and developed by Dowker, Wallden, Ghazi-Tabatabai, and Wilkes. We now present its basic ideas following the line of Wallden (2013) and analyze some ontological issues therein.

The construction of this formalism is based on the quantum histories, events, and the corresponding measures. The reader may find it helpful to review the relevant formulas for the decoherence functional and the probability of history introduced

DOI: 10.4324/9781003662860-8

in Chapter 3. In quantum measure theory, the quantum measure of a quantum history is also defined as

$$\mu(A) := D(A, A) = Tr\left(\hat{C}_A^{\dagger}\rho\hat{C}_A\right)$$

where $\hat{C}_A$ is the chain operator of the history and ρ represents the initial condition information. For cases with an infinite number of continuous moments or time variables that are difficult to clearly define, the chain operator can be replaced by the Feynman history in the form of path integral:

$$\mu(A) = \left| \int_{A \subseteq \Omega} \exp\left(iS\left[x(t)\right]\right) \mathcal{D}x(t) \right|^2$$

where A is a subset of the entire event set Ω. By observing this definition and those formula in Chapter 3, it can be confirmed that the quantum measure of an event A is exactly the formula for the occurrence probability of a quantum history A in a specific history family.[1]

However, for general quantum history families, if the decoherence or consistency condition or other constraint conditions are not met, there may be interference between two histories, and thus the quantum measure does not satisfy the property of additivity. Therefore, generally speaking, the quantum measure should not be directly understood as probability. That is, in quantum measure theory,

$$\mu(A \sqcup B) \neq \mu(A) + \mu(B)$$

But the quantum measure defined as above satisfies positivity,

$$\mu(A) \geq 0 \text{ for all } A \in \Omega$$

normalization,

$$\mu(\Omega) = 1$$

and the following condition:

$$\mu(A \sqcup B \sqcup C) = \mu(A \sqcup B) + \mu(B \sqcup C) + \mu(A \sqcup C) - \mu(A) - \mu(B) - \mu(A) \qquad (8.1)$$

Note that this is different from the usual union operation:

$$A \cup B \cup C = (A \cup B) + (B \cup C) + (A \cup C) - (A \cap B) - (B \cap C) - (A \cap C) + (A \cap B \cap C)$$

This condition means that the quantum measure is fully determined once we know the quantum measure of single histories and pairs of histories. Formula (8.1) has been experimentally verified (Sinha et al., 2010). With the definition of quantum measure, we can construct a theory that characterizes quantum measure. We need the following triplet: the histories space, the events algebra, and the quantum measure $(\Omega, \mathcal{U}, \mu)$.

A partition $\mathcal{P}$ of the histories space Ω is a collection of events $\{P_1, P_2, \cdots\}$ such that $P_i \cap P_j = \varnothing$ for all $i \neq j$ and $U_i P_i = \Omega$. It is an exhaustive and exclusive collection. Each event of the partition is called cell. A coarse-graining of a histories triplet $(\mathcal{P}, \mathcal{U}_\mathcal{P}, \mu_\mathcal{P})$ is defined in the following way (Wallden, 2013: 6):

1 The cells of a partition $\mathcal{P}$ are the elements of the coarse-grained histories space.
2 The Boolean algebra generated by the addition and multiplication of these cells $\mathcal{U}_\mathcal{P}$ is the coarse-grained event algebra.
3 The quantum measure defined on them should be compatible with the fine-grained quantum measure, i.e., $\mu_\mathcal{P}(A) = \mu(A)$ for all $A \in \mathcal{U}_\mathcal{P}$.

The new ontology of quantum theory is the coevent, which is not affected by measurement (unlike the Copenhagen quantum mechanics) and is not contextual. In this sense, the coevent formalism is realist. We explain the ontology of the coevent formalism in more detail. There are two equivalent perspectives for the ontology of this formalism, which are in actual dual to each other. Specifically, the first holds that reality (the ontology of the theory) is a coarse-grained quantum history. The second view holds that reality is a (generalized) valuation mapping φ, which is usually non-homomorphic. In classical physics, we can also consider reality as a homomorphic mapping φ_{h_i}, which is the characteristic function of the classical histories h_i (please compare with the superdensity operators formalism discussed later).

The second view presupposes a basic idea that the characteristic function mapping reality to truth values can itself be regarded as reality:

$$\varphi_h(A) = 1 \text{ iff } h \in A$$

Here, h is a history, and A is an event in h. The coevent is a generalization of this idea. As Sorkin put it,

> we can also think of an event as a question.... A coevent can then be thought of as something that answers all conceivable questions ... in some sense a 'higher-order object.' If one thought of events as 'predicates' then a coevent would be a 'predicate of predicates'.
>
> (Sorkin, 2012: 365–366)

A truth valuation mapping from events/histories to truth values, which will be shown below, is also an anhomomorphic valuation. The generalization from the classical case to the quantum case is precisely the modification that weakens the requirement that the truth valuation mappings are homomorphisms between the Boolean algebras of $\mathcal{U}$ and $\mathcal{T}$. As anhomomorphisms, they are multiplicative but no longer additive.

The operation rules that the coevent φ satisfies are

$$\varphi(A\,B) = \varphi(A \cap B) = \varphi(A)\varphi(B)$$

$$\varphi(A + B) \neq \varphi(A) + \varphi(B)$$

For the coevent φ_S, if $\varphi_S(A) = 1$, it indicates that A is an event in S, which exactly agree with the definition of a characteristic function. Therefore, all sets that can make φ take the value of 1 are events. In other words, events are the support of the coevent.

Two additional constraints are also required. First, they are preclusive:

$$\mu(P) = 0 \Rightarrow \varphi(P) = 0$$

That is, for events with a measure of 0 (precluded events), the coevent gives the answer "false". Second, the coevent φ should be a primitive preclusive multiplicative map, that is, there does not exist another preclusive multiplicative mapping ψ such that:

$$\varphi(A) = 1 \Rightarrow \psi(A) = 1 \ \forall A \in \mathcal{U}$$

In summary, mappings that satisfy all the above definitions and constraints are called coevents, which are the ontology of this theory. Although it is away from classical physics intuitive, the following points can demonstrate its reality. (1) With a natural modification to the above definition, such as requiring it to be an additive homomorphism, it can be fully applicable in the classical physics case. The above requirements for primitiveness precisely make the coevents in the classical case the characteristic mappings of single histories. (2) Coevents are noncontextual. They are not affected by measurements and can apply to closed quantum systems, which means that it is well well-suited for the study of quantum cosmology.

However, the deviation from the classical reality lies in that, "the quantum measure allows the existence of zero covers. We therefore can no longer maintain the picture of reality as a single fine-grained history or equivalently as a homomorphic map" (Wallden, 2013: 12). Instead, the picture of reality contains coarse-grained histories or multiplicative mappings). After this generalization, we have potential realities that are compatible with any quantum measure, which contain information about initial conditions and dynamics. In the coevent formalism, a peculiar phenomenon is that the exact form of the logic here – that is, when other inference rules can be applied – depends on the possible coevent sets, and these coevent sets, in turn, depend on the specific quantum measure. Therefore, the reality is realist, but the logic of the quantum world is dynamical. To quote Sorkin, one of his famous article has an unequivocal title as "Logic is to quantum, as geometry is to gravity" (Sorkin, 2012). Please recall the discussion and remarks in Section 6.3.

Compare it with the quantum histories interpretation. The quantum histories approach utilizes the decoherence functional to find a partition $\mathcal{P}_i = \left\{P_1^i, P_2^i, \cdots\right\}$ of the history space Ω such that there is no interference between any pair of cells in the partition (satisfying certain selection conditions). According to Griffiths, the quantum measure (probability assignment method) induced in this way becomes a classical measure (probability). Each such partition is called a consistent set, and each consistent set is regarded as a classical domain. As described in Chapters 3 and 4, there does not exist a most fine-grained consistent set from which all other consistent sets can be derived. Properties within a consistent set may be incompatible with those in another consistent set. Either a preferred consistent set is selected by some physical mechanism (but currently unknown), or, as Griffiths suggests, unicity is abandoned and the answer to a quantum proposition depends on the consistent set being referred to. Quantum histories interpretation is therefore an A-type theory in the Penrose sense of needing to be supplemented by a theory of experience (Dowker and Wilkes, 2022: 11).

Some philosophical questions that need further research include (Wallden, 2013: 17): From the coevent itself only, which properties of the system can we infer? What properties does the coevent itself possess? How to formally extend the quantum measure theory to the infinite histories space? How to deal with the composite system? How to extend the time division of the coevent (the influence of the fine-graining of the quantum measure on the allowed coevent set)? How about the large systems where classicality should emerge?

The second type of generalization is the superdensity operator formalism, the main idea of which is to put space and time in a similar position in quantum theory, so that the formalism can be applied to the problems of quantum information and entanglement dynamics in spacetime (Cotler et al., 2017). This deviates from the usual quantum mechanics formalism, where people treat space and time in a completely different way. From this perspective, if one wants to make time no longer just a parameter in the single-time description, one has to consider the extended Hilbert space, which includes the tensor product of Hilbert spaces at different times:

$$\mathcal{H}_{\text{hist}} = \otimes_{\boldsymbol{x},t} \mathcal{H}_{\boldsymbol{x},t}$$

This is just the history of Hilbert space proposed by Griffiths. In the consistent histories framework and its variants, the core object is the chain operator composed of the projection operators on $\mathcal{H}_{\text{hist}}$. The entangled histories formalism developed by Cotler and his collaborators generalizes the consistent histories formalism to define quantities more similar to spacetime states (allowing superposition) (see Cotler and Wilczek, 2015; Cotler and Wilczek, 2016; Cotler and Wilczek, 2017; Cotler et al., 2017; Nowakowski, 2017; Milz et al., 2021; also see Brukner et al., 2004). From a mathematical perspective, the superdensity operator is a quadratic form on the space of operators on the history Hilbert space. From a physical perspective, the superdensity operator of a physical system encodes its response to experimental detection, allowing these detections to occur at different times (and locations). More abstractly, just as the standard density operator represents a state and encodes the data of all correlation functions at a fixed time, we propose that the superdensity operator represents a spacetime state and encodes the data of all spacetime correlation functions. As its name suggests, this kind of operator shares many similar properties with the density operators. The entanglement, entropy, mutual information, etc., of the superdensity operator are all operationally meaningful physical concepts, and thus of great significance for the definition of the spacetime state of a system.

A superdensity operator ϱ is defined as a bilinear form on the history Hilbert space $\mathcal{H}_{\text{hist}}$:

$$\varrho : B^*\left(\mathcal{H}_{\text{hist}}\right) \otimes B\left(\mathcal{H}_{\text{hist}}\right) \mapsto \mathbb{C}$$

$B\left(\mathcal{H}_{\text{hist}}\right) := B\left(\mathcal{H}_{t_1}\right) \odot B\left(\mathcal{H}_{t_2}\right) \odot \cdots \odot B\left(\mathcal{H}_{t_n}\right)$ denotes the space of bounded operators on $\mathcal{H}_{\text{hist}} = \mathcal{H}_{t_1} \odot \cdots \odot \mathcal{H}_{t_n}$. It satisfies the following conditions:

1 Hermitian: $\varrho^\dagger = \varrho$;
2 Positive semi-definite: $\varrho \succeq 0$, meaning $\langle W | \varrho | W \rangle \geq 0$ for all $|W\rangle$;
3 Unit trace: $tr(\varrho) = 1$.

Superdensity operators have a close connection with density operators. In contrast to the definition of density operators, a density operator ρ is defined as a bilinear form on the Hilbert space $\mathcal{H}$ (say, at a certain time t_d):

$$\rho : \mathcal{H}^* \otimes \mathcal{H} \mapsto \mathbb{C}$$

satisfying the conditions:

1 Hermitian: $\rho^\dagger = \rho$;
2 Positive semi-definite: $\rho \succeq 0$, meaning $\langle \phi | \rho | \phi \rangle \geq 0$ for all $|\phi\rangle$;
3 Unit trace: $Tr(\rho) = 1$.

Similar to Dirac's bra–ket notation, Cotler et al. introduced a soft bracket notation to represent operators in $B\left(\mathcal{H}_{\text{hist}}\right)$ and $B^*\left(\mathcal{H}_{\text{hist}}\right)$. Operators in $B\left(\mathcal{H}_{\text{hist}}\right)$ are denoted as $\left|W\right)$, and the dual operator in $B^*\left(\mathcal{H}_{\text{hist}}\right)$ corresponding to $\left|W\right)$ is denoted as $\llcorner \left(V\right| := Tr\left(V^\dagger \cdots\right)$, where the ellipsis "…" will be filled with $\left|W\right)$. They satisfy linearity relations and multilinear relations for tensor products.

Therefore, a typical example of a superdensity operator can be expressed as

$$\varrho = \frac{1}{d^2} \sum_{\substack{i_1,\ldots,i_n \\ j_1,\ldots,j_n}} Tr\left(X_{i_n} U\left(t_n, t_{n-1}\right) \cdots U\left(t_2, t_1\right) X_{i_1} \rho_0 X_{j_1}^\dagger U\left(t_1, t_2\right) \cdots U\left(t_{n-1}, t_n\right) X_{j_n}^\dagger\right)$$

$$\times \left|X_{i_1}\right)\left(X_{j_1}\right| \otimes \cdots \otimes \left|X_{i_n}\right)\left(X_{j_n}\right|$$

Next, the trace operation and partial trace operation for the density operator can be extended to superdensity operator:

$$Tr(\mathcal{S}) = \sum_{i=1}^{\dim B(\mathcal{H}) = (\dim(\mathcal{H}))^2} (X_i|\mathcal{S}|X_i)$$

$$Tr_A(\mathcal{S}) = \sum_{i=1}^{\dim(A)} \left((Y_i|_A \otimes \mathbf{1}_B) \mathcal{S} (|Y_i)_A \otimes \mathbf{1}_B \right)$$

From another perspective, a bilinear map

$$\varrho : B^*(\mathcal{H}_{\text{hist}}) \otimes B(\mathcal{H}_{\text{hist}}) \mapsto \mathbb{C}$$

can also be regarded as a superoperator:

$$\varrho : B^*(\mathcal{H}_{\text{hist}}) \mapsto B(\mathcal{H}_{\text{hist}})$$

The relationship between consistent histories and the superdensity operator formalism can be briefly summarized as follows. In the language of superdensity operators, we restrict the superdensity operator to projectors on the history Hilbert space (rather than all bounded operators):

$$\tilde{\varrho} : \text{Proj}^*(\mathcal{H}_{\text{hist}}) \otimes \text{Proj}(\mathcal{H}_{\text{hist}}) \mapsto \mathbb{C}$$

where $\text{Proj}(\mathcal{H}_{\text{hist}})$ is the space of projectors on $\mathcal{H}_{\text{hist}}$, and $\text{Proj}^*(\mathcal{H}_{\text{hist}})$ is its duality. If there is a series of projectors satisfying $\Sigma_i |P_i) = \mathbf{1}_{\mathcal{H}_{\text{hist}}}$, then the consistency condition can be expressed as

$$(P_i|\tilde{\varrho}|P_j) = \delta_{ij}(P_i|\tilde{\varrho}|P_i)$$

Note that $\text{Proj}(\mathcal{H}_{\text{hist}})$ does not simply form a linear subspace of $B(\mathcal{H}_{\text{hist}})$. This restricted partial diagonalization of $\tilde{\varrho}$ is not unique, and thus there are many possible consistent families, which is a conclusion we have already known from the previous chapters.

Conceptually, the superdensity operator and the consistent histories formalism are "dual" – the superdensity operator maps histories to numerical values. This means that one can use the basic elements of quantum information theory in the superdensity operator formalism (such as partial traces and entropy) to define corresponding meaningful notions for histories. This is more general than the quantum histories formalism because the superdensity operator can accept any operators as input, not just projectors, thus flexible enough to grasp quite a lot of irregular situations from quantum information and quantum gravity theory. In this sense, it is a radical generalization of the quantum histories formalism. Histories and more generally entangled histories (higher rank projectors in $\mathrm{Proj}(\mathcal{H}_{\mathrm{hist}})$) are in fact observable by coupling to superdensity operator to auxiliary apparatus (Cotler et al., 2018: 45–46).

The third generalization is called the pseudo-density matrix formalism (Zhang, Dahlsten, and Vedral, 2020: 2–3). This formalism is originally proposed as a finite-dimensional quantum-mechanical formalism that aims to treat space and time on an equal footing but can also be generalized to the infinite and continuous cases. Event is defined via making a measurement in spacetime and is built upon correlations from measurement results. Thus, temporal correlations are treated just as spatial correlations from observation of measurements, and spatiotemporal correlations are unified in a single framework.

Consider a n-qubit density matrix, it can be expanded by Pauli operators σ_i in terms of Pauli correlations. In spacetime, pick up n events; for each event, a single-qubit Pauli operator is measured. The pseudo-density matrix in the finite-dimensional case is then defined as

$$\hat{R} \equiv \frac{1}{2^n} \sum_{i_1=0}^{3} \cdots \sum_{i_n=0}^{3} \left\langle \left\{ \sigma_{i_j} \right\}_{j=1}^{n} \right\rangle \otimes_{j=1}^{n} \sigma_{i_j}$$

$\left\langle \left\{ \sigma_{i_j} \right\}_{j=1}^{n} \right\rangle$ is the expectation value of the product of the measurement results for events with measurement operators $\left\{ \sigma_{i_j} \right\}_{j=1}^{n}$. For spacelike separated measurements or independently evolving local systems, the pseudo-density matrix will reduce to a standard density matrix. But for measurements made in time, the pseudo-density matrix may have negative eigenvalues.

The relation with the n-qubit pseudo-density matrix is as follows. An n-qubit pseudo-density matrix can be considered as a single qubit evolving at n times. For each event, make a single-qubit Pauli measurement σ_{ik} at the time t_k. Separate the measurement σ_{ik} into two projection operators $P_{i_k}^{+1} = \frac{1}{2}\left(I + \sigma_{ik}\right)$ and $P_{i_k}^{-1} = \frac{1}{2}\left(I - \sigma_{ik}\right)$ with its outcomes ± 1. Each pseudo-density event with the measurement σ_{ik} corresponds to two history events with projections $P_{i_k}^{\pm 1}$. A pseudo-density matrix is built upon measurement correlations $\left\langle \left\{ \sigma_{ik} \right\}_{k=1}^{n} \right\rangle$. In terms of decoherence functionals, that is

$$\left\langle \left\{ \sigma_{ik} \right\}_{k=1}^{n} \right\rangle = \sum_{\alpha_1,\ldots,\alpha_n} \alpha_1 \ldots \alpha_n Tr\left[P_{i_n}^{\alpha_n} U_{n-1} \cdots U_1 P_{i_1}^{\alpha_1} \rho P_{i_1}^{\alpha_1} U_{n-1}^{\dagger} \cdots U_1^{\dagger} P_{i_n}^{\alpha_n} \right]$$

$$= \sum_{\alpha_1,\ldots,\alpha_n} \alpha_1 \ldots \alpha_n p\left(\alpha_1 \ldots \alpha_n\right)$$

$$= \sum_{\alpha_1,\ldots,\alpha_n} \alpha_1 \ldots \alpha_n D([\alpha],[\alpha])$$

where $D([\alpha],[\alpha])$ is the diagonal terms of decoherence functional with $[\alpha]=[\alpha_1,\ldots,\alpha_n]$. Only diagonal decoherence functionals are taken into account, which amounts to the decoherence condition. As a result, the diagonal terms of decoherence functionals in quantum histories are the probabilities in temporal correlations of pseudo-density matrices. The decoherence condition also explains why spacetime states for general measurements are defined in diagonal form.

It is worth adding here that, for the historicized ontology of quantum mechanics, these three formalisms all seem to have a dualistic sense. It may be like the relationship between homology and cohomology in the homology theory of algebraic topology, the dual structures/operations may show asymmetrically greater power and occupy a higher-level position, and thus can immeasurably enlarge the original theoretical expressiveness. However, for the duality, or measure, of the historicized ontology, the physical meanings of them will be more abstract, and the specific process of calculation and computation of them seems extremely complicated. The complexity has greatly increased in calculating concrete cases compared with the quantum histories, but whether this has any ontological significance and implications is also worth further consideration.

8.2 Possible future research directions – Part II: various physical and philosophical aspects

The new concepts proposed by the quantum histories interpretation are quite different from the concepts in classical physics and also from the basic pictures given by other interpretations of quantum mechanics, which have sparked many controversies and criticisms. The two main issues are about ontology and logic. This book mainly discusses the ontological problems of discontinuity in the assignment of properties, the difficulty in maintaining quasi-classicality, the relevance of reality to the framework, the dependence of initial states and consistency conditions on the framework, the incomplete solution to the measurement problem, and the lack of explanation for the quasi-classical experience existing in the macroscopic world, as well as the logical problems of contradictory results when making predictions and retrospections from known consistent history families, the derivation of mutually contradictory conclusions in the probabilistic proposition reasoning across history families, and the contradiction in the truth value assignment of the same proposition in two compatible frameworks when there is a contextual correlation of observable quantities. It points out that the emergence of these two types of problems and their correlation are related to the special formalism of the quantum histories interpretation.

Subsequently, this book analyzes the discussion of the primitive ontology that emerged from Bohmian mechanics and the nomological interpretation of the wave function of quantum mechanics, especially the concept of "nomological-ity" in different senses, including universal nomology, quasi-nomology, strong nomology, and weak nomology, and evaluates the related doubts and responses. This nomological interpretation brings important inspirations to our examination of the formalism of quantum histories and their interpretation problems and helps us understand the ontological and logical difficulties therein. According to the terminological definitions in the previous chapters and the discussion of various selection conditions for history families of different strengths, this book finally holds that the corresponding quantum histories formalisms are not yet complete theories, and their ontological commitments and nomological relations have not been clearly presented. Therefore, they themselves still need to be continuously developed and interpreted. Each selection method for history families actually corresponds to a quantum histories formalism, and each such formalism has not yet truly completed its own interpretation (except for the consistent history formalism, which currently has a relativized realm-related reality interpretation proposed by Griffiths himself). If each history in the history family that meets the selection conditions is regarded as the most fundamental reality of the theoretical presupposition, and various selection methods are regarded as nomological, this perspective does have some rationality, but it also faces many difficulties. In addition, if it is believed that there is only one quasi-classical framework in which we live in this world, then obviously all the above selection methods are still unable to pick it out, and thus are not "optimal". There is even reasonable doubt about whether there can be a so-called "optimal" selection method. In this sense, the research on quantum histories formalisms is still in an unfinished state. The "interpretation" and "theoretical extension" of existing and future quantum histories formalisms need to make substantial progress through their interactive promotion.

One important research direction in the future is to deeply understand the nature of the wide spectrum of known and unknown selection conditions. It can be seen that each selection condition has its own advantages and disadvantages when selecting the admissible history families that are considered to have good properties: stronger conditions can satisfy good geometric properties, dynamic stability, consistency for composite systems and subsystems, etc., but they possibly exclude some actual quantum histories; weaker conditions have a wiper accommodation, but they also lose reasonable constraints to exclude huge amount of nonsense histories. Besides, people have not yet found a general method to gain the quasi-classical realms that Hartle and others have envisioned in specific cases, nor have they found the selection conditions that can exactly define the quasi-classical realms. If such selection conditions exist, they must be stronger than any known conditions, such as consistency conditions and decoherence conditions, and involve the properties of multi-particle systems and macroscopic scales. Whether they present more special properties and whether they are very complex (or even unsolvable) in computational solutions, we are still not clear. Some scholars believe that spontaneous collapse dynamics should be introduced into the quantum histories formalism (Kent, 1997) to find the expected quasi-classical history families, but this approach seems to shift the problem to the difficulties faced by various spontaneous collapse theories.

Most of the current quantum histories interpretation literature discusses how to understand the concepts and quantum paradoxes of non-relativistic, closed systems in quantum mechanics. Therefore, another important research direction is to extend the non-relativistic quantum histories formalism to relativistic quantum mechanics, quantum field theory, quantum cosmology, and open quantum systems. As Kent said:

> If the eventual goal is to go beyond quantum theory, it is probably as well to understand all interesting formulations and interpretations of the theory. Different formulations, after all, may inspire different post-quantum theories. We need, in particular, to understand … a general treatment of quantum field theory in the consistent histories formulation; and a clearer understanding of how the formalism applies to cosmology.
>
> (Kent, 1996b: 352)

Research in this direction is currently only in its rudimentary stage, such as the related work of Hartle and Gell-Mann, the discussion of quantum cosmology and the measurement problem in it by Craig and Singh (2011), Halliwell's analysis of the probability in cosmological models using decoherent histories (Halliwell, 2009b), the discussion of quantum field theory in curved spacetime (Anastopoulos, 2002; Blencowe, 1991), the discussion of candidate theories for quantum gravity and quantum cosmology (such as loop quantum cosmology, Schroeren, 2013), the discussion of quantum geometric phases (Anastopoulos and Savvidou, 2002), and the discussion of quantum information (Griffiths, 2017). In addition, some scholars have attempted to rephrase classical physics using the method of history projection operators, including classical electrodynamics and general relativity (Burch, 2004; Savvidou, 2001). Currently, most of these are only theoretical and mathematical research, lacking responses to philosophical concerns. Due to the serious fundamental concept interpretation problems in subjects such as quantum field theory and cosmology, they may even be more disruptive than those in non-relativistic quantum mechanics (Kronz and Tracy, 2025; Kuhlmann, 2023), whether the understanding of quantum histories from the perspective of nomology presented in this book can be reasonably extended to these theories still needs to be discussed in combination with the progress of the interpretation work on these theories. Besides, when it comes to deal with the "arrival time problem" (Halliwell and Zafiris, 1998; Halliwell and Yearsley, 2009; Halliwell, 2009b; Yearsley, 2010), the time-dependent entanglement phenomenon (Cotler and Wilczek, 2016; Cotler et al., 2017), and the time issue in cosmology and quantum gravity (Isham, 1993; Christodoulakis and Wallden, 2011; Anderson, 2017), the approach of quantum histories formalism can demonstrate special advantages.

The third research direction is to continue to improve and expand the quantum histories interpretation, as well as to develop other interpretations borrowing the basic elements from quantum histories formalism, for example, to expand the quantum histories on discrete moments to those on continuous time (Isham et al., 1998; Anastopoulos, 2001; Savvidou, 1999a, 1999b), the history across the framework and expanded definition of probability within the "regular histories interpretation" (Priebe, 2012), and further extend the quantum measure theory and the coevents formalism (Ghazi-Tabatabai, 2009, 2013; Gudder, 2011; Sorkin, 2012; Wallden, 2013) and handle relativistic causality through the behaviors of quantum events with zero measure (Dowker and Sorkin, 2023), the "superdensity operator" formalism and its application to consider quantum states which encode spacetime dynamics, and develop quantum information theoretic tools to extract the causal relationships between putative spacetime subsystems (Cotler et al., 2019), the Events-Trees-Histories interpretation (Fröhlich, 2020, 2021, 2024), the tensor networks for quantum causal histories (Guo, 2020), etc. These extended versions may offer ideas for solving the inherent difficulties of the quantum histories formalism. Some of them, if they can be developed further, might even trigger the new transformation of the ontological picture implied by the quantum histories formalism.

The fourth direction is to examine the connection between the quantum histories interpretation and other interpretations of quantum mechanics. For instance, Hartle and Gell-Mann once discussed the relationship between the quantum histories interpretation and the Copenhagen interpretation, Bohm mechanics, and Feynman's sum-over-histories formulation, and they believed that the latter two could be regarded as special cases of the quantum histories description of closed systems under certain restrictions (Hartle, 2002, 2004); Rovelli once discussed the connection between the quantum histories interpretation and the relational interpretation, arguing that there is no fundamental conflict between the two, and that the relational interpretation can provide a useful supplement to the quantum histories interpretation (Rovelli, 1996: 1673–1674), thus potentially combining the quantum histories interpretation's emphasis on the consistency of event chronology and the proposition framework to supplement and develop the relational interpretation and solve the problem of information sharing among multiple observers (Robson, 2024); some have attempted to understand the quantum histories from the perspective of Bayesian probability (Marlow, 2006), which might provide a reference for the future combination of the quantum Bayesian interpretation and the quantum histories interpretation; some scholars have considered introducing the basic ideas of quantum histories into the formulation of the modal interpretation (Hemmo, 1996, 1998); in the recent years, some scholars have re-examined the relationship between the quantum histories interpretation and the many-worlds interpretation (Mandolesi, 2017; Bigaj, 2023). In addition, for the extension of the quantum histories formulation in fields such as quantum field theory and quantum cosmology, more comparative studies can be conducted with the more commonly applied Feynman's sum-over-histories formulation in these fields. By carefully examining the similarities and differences in conceptual meanings between other interpretations and the quantum histories interpretation through more concrete case studies of different applications (especially in situations involving the early universe and even open quantum systems, etc.), perhaps some unexpected insights can be gained into the inherent commonalities of these interpretations, thereby deepening the understanding of the essence of quantum mechanics. A notable attempt in this regard in recent years is mentioned in Chapter 5, where some philosophers of science who support Humeanism aim to unify the understanding of the ontology and nomology in Bohm mechanics, spontaneous collapse theories, and even certain versions of the many-worlds interpretation from the primitive ontology. From the Humeanism perspective, can a broader, unified understanding of quantum mechanics interpretations be achieved? This might be a direction worth exploring.

The fifth research direction is to reflect on the common understanding of basic concepts, such as the law of nature, in the philosophy of science, through the nomological interpretation of the quantum histories formalism. The majority of the current literature about the essence of the law of nature takes the "laws" concerned in classical mechanics, thermodynamics, electromagnetism, optics, chemistry, biology, and even psychology and economics as examples for discussion, while less attention is paid to whether the "formulas", "equations", and "principles" in more abstract modern physics (such as condensed matter physics and high-energy physics) are also applicable to the several main views on the law of nature. On the one hand, the philosophers of science and even scientists themselves have not yet clearly understood the ontological basis of these theories with more abstract and higher-order mathematical expressions (for instance, the mathematical construction of some theories has not been rigorously formalized, and the framework of some theories is still not clear). Under such circumstances, in-depth discussions on laws of nature may require more preparatory work than before. On the other hand, only by reviewing previous opinions on the law of nature in combination with more cutting-edge progress in physical science, can we consider the law of nature and related issues such as natural kinds and scientific explanations from a new perspective. In fact, it may not be limited to physics, but also needs to be combined with materials from rapidly developing disciplines, including mathematical biology, cognitive science, and evolutionary game theory.

Of course, given that the quantum histories interpretation provides us with profound physical ideas and philosophical implications, potential future research directions are not confined to those outlined above. In the future, researchers will undoubtedly discover additional valuable questions and may even pioneer entirely new and significant fields of exploration. The research initiated by the quantum histories interpretation merits further serious investigation by physicists together with philosophers of science.

8.3 Concluding remarks from the methodological considerations for future foundational research of quantum mechanics

Finally, with respect to the prospects of the inherent issues in the controversial interpretations of quantum mechanics, the author would like to offer the following tentative considerations. What precisely gives rise to the difficulties in interpreting quantum mechanics? Currently, at least the following aspects are worthy of reflection.

First, if physicists maintain the belief that uncovering a unified underlying theory or some "Theory of Everything" is desirable and achievable, then quantum mechanics, which aims to describe physical entities above the Planck scale, can merely be an effective theory rather than a fundamental or underlying one. As Michael Heller put it,

> If the present quantum mechanics is incomplete then it is cutoff from a larger structure, more or less as the top of an iceberg is cut off, for our eyes, from what is hidden in the water. It is quite natural that the strange behavior of the visible part cannot be fully understood without taking into account the behavior of the rest.
>
> (Grygiel, 2007: 97)

The yet-to-be-developed deeper theory might render quantum mechanics an approximation to a certain limit. Consequently, the various conundrums and paradoxes within quantum mechanics could potentially be explicated in such a deeper theory, providing a conclusive outcome to the long-standing debate over the interpretation of quantum mechanics and fundamentally resolving its non-classical peculiarities, such as incompleteness, non-locality, indeterminacy, and contextuality. This deeper answer might not be unique, but it ought to present a clearer and simpler picture than the current chaotic state, where numerous interpretations coexist and none can be readily falsified or verified, and attain a clarity comparable to our present understanding of classical physics.

On the other hand, even if this were the case, it does not imply that the fundamental inquiry has come to an end. Even if in the future people were to develop an experimentally confirmed, relatively complete underlying theory (by relatively complete, it is meant roughly equivalent to what is currently known as classical mechanics and non-relativistic quantum mechanics), such as a certain version of the ultimate theory or the grand unified theory, new interpretive difficulties and paradoxes would still emerge. Lee Smolin, a researcher in quantum gravity, contends that the supplementation or completion of quantum mechanics by a deeper theory "differs from interpretation", signifying that "new dynamics need to be provided" (Smolin, 2021: 2). Hence, the basic variables or ontology of this new dynamics, whether they are the various iterations of hidden variable theories in the early days or the "views" and "varieties" based on causal set structures introduced by Smolin and others recently, will engender new interpretive problems – will these new basic variables be more lucid and comprehensible than the quantum states that are so distinct from classical physical reality? Generally, in the theoretical explanation of physical experimental results, "although more complicated methods and a greater number of parameters typically lead to better predictions, when assessing the quality of a theory, the relationship between the number of assumptions and the predictive power must be balanced" (Sun, 2024: 48); we should pursue a simple and comprehensible image of the world in the appropriate manner as Einstein had always desired. Similar predicaments also exist in the philosophical interpretation of theories, where one needs to strike a balance among the complexity of the ontological presuppositions of the interpretation, the simplicity and consistency of the interpretive assumptions, and the interpretive power ultimately manifested.

Perhaps, the pursuit of an underlying theory is not the sole means to resolve the puzzles and paradoxes of quantum mechanics. In this less optimistic situation, different possibilities may emerge:

1 The puzzles and paradoxes of quantum mechanics exhibit "heredity" and "intrinsicality", meaning that any deeper theory capable of accommodating the formalism of quantum mechanics and treating it as a low-energy effective approximation will, in some fashion, "inherit" these formidable challenges. Of course, these puzzles and paradoxes will then be represented through distinct formalisms with entirely novel mathematical structures and even new logic.
2 The main puzzles and paradoxes within quantum mechanics may be resolved therein, but this deeper-level, higher-energy theory will give rise to more unforeseen difficulties.
3 Irrespective of the energy scale, these puzzles and paradoxes that arise at the quantum level or microscopic scale might even prove insoluble in principle from within the domain of physics. It is imaginable that the increasingly fundamental theories we discover remain effective approximate theories, confronted with the interpretive challenges that effective theories cannot evade, and a fully-fledged grand unified theory can never be attainable. If such is the case, it seemingly intimates that human beings will reach the boundary of their cognitive capacity in physics or the unknowable limits of the deep material world. The ultimate question then will be like: if there exists a boundary to our cognition of physics/the physical world, does this still fall within the realm that physics is capable of investigating, can it still be resolved in any way within the science of physics, or has it even surpassed the conceivable potential of human rationality? If the scientific interpretation at the intersection of the humanities and the exact sciences harbors the aforementioned uncertainties, then its essence is undoubtedly an interrogation to the boundaries of human rationality. In the process of interpretation, the enterprise of science, as an embodiment of human rationality, continuously reveals its openness and infinitude within its finiteness. That echoes what Whitehead has told us more than 80 years ago:

> The notion of 'understanding' requires some grasp of how the finitude of the entity in question requires infinity, and also some notion of how infinity requires finitude. This search for such understanding is the definition of philosophy.
>
> (Whitehead, 1941: 675)

Note

1 However, Dowker and Wilkes admitted that "the relationship between the quantum measure and the decoherence functional and their physical significance – including the question of which is the more primitive concept – remain to be fully worked out" (Dowker and Wilkes, 2022: 3).

References

Acuña P., 2019, Charting the landscape of interpretation, theory rivalry, and underdetermination in quantum mechanics, *Synthese*, 198, 1711–1740.

Albert D.Z., 1992, *Quantum Mechanics and Experience*, Harvard University Press.

Allori V., 2013, On the metaphysics of quantum mechanics, in Lebihan S. ed., *Precis de la Philosophie de la Physique*, pp. 116–151, Vuibert.

Allori V., 2015, Primitive ontology in a nutshell, *International Journal of Quantum Foundations*, 1(3), 107–122.

Allori V., 2017, A new argument for the nomological interpretation of the wave function: The Galilean group and the classical limit of nonrelativistic quantum mechanics, *International Studies in the Philosophy of Science*, 31(2), 177–188. https://doi.org/10.1080/02698595.2018.1424759

Allori V., 2019, Quantum mechanics, time and ontology, *Studies in History and Philosophy of Science Part B: Studies in History and Philosophy of Modern Physics*, 66, 145–154. https://doi.org/10.1016/j.shpsb.2019.02.004

Allori V., 2020, Scientific realism without the wave function, in French S. and Saatsi J., eds., *Scientific Realism and the Quantum*, pp. 212–228, Oxford University Press.

Allori V., Goldstein S., Tumulka R. and Zanghì N., 2008, On the common structure of Bohmian mechanics and the Ghirardi-Rimini-Weber theory: Dedicated to Giancarlo Ghirardi on the occasion of his 70th birthday, *The British Journal for the Philosophy of Science*, 59(3), 353–389.

Allori V., Goldstein S., Tumulka R. and Zanghì N., 2011, Many-worlds and Schrödinger's first quantum theory, *British Journal for the Philosophy of Science*, 62(1), 1–27.

Allori V., Goldstein S., Tumulka R. and Zanghì N., 2014, Predictions and primitive ontology in quantum foundations: A study of examples, *The British Journal for the Philosophy of Science*, 65(2), 323–352.

Anastopoulos C., 2001, Continuous-time histories: observables, probabilities, phase space structure and the classical limit, *Journal of Mathematical Physics*, 42(8), 3225–3259.

Anastopoulos C., 2002, Histories quantization for quantum fields in curved spacetime, in Gurzadyan, V.G., Jantzen, R.T. and Ruffini, R., eds., *The Ninth Marcel Grossmann Meeting: On Recent Developments in Theoretical and Experimental General Relativity, Gravitation and Relativistic Field Theories* (3 Volumes), pp. 1363–1364, World Scientific.

Anastopoulos C. and Savvidou N., 2002, Quantum mechanical histories and the Berry phase, *International Journal of Theoretical Physics*, 41(3), 529–540.

Anderson E.J., 2017, *The Problem of Time: Quantum Mechanics Versus General Relativity*, Springer.

Bassi A. and Ghirardi G., 1999, Can the decoherent histories description of reality be considered satisfactory? *Physics Letters A*, 257(5–6), 247–263.

Bell J.S., 1987, *Speakable and Unspeakable in Quantum Mechanics: Collected Papers on Quantum Philosophy*, Cambridge University Press.

Bigaj T., 2023, Consistent histories and many worlds, *Quantum Reports*, 5, 186–197.

Bhogal H. and Perry Z., 2017, What the humean should say about entanglement, *Noûs*, 51(1), 74–94.

Blencowe M., 1991, The consistent histories interpretation of quantum fields in curved spacetime, *Annals of Physics*, 211(1), 87–111.

Boës P. and Navascués, M., 2017, Composing decoherence functionals. *Physical Review A*, 95(2), 022114.

Bohm D., 1952, A suggested interpretation of the quantum mechanics in terms of hidden variables: I & II, *Physical Review*, 85, 166–179, 180–193.

Brody D.C. and Hughston L. P, 2001, Geometric quantum mechanics, *Journal of geometry and physics*, 38(1), 19–53.

Brukner Č., Taylor S., Cheung S. and Vedral V., 2004, *Quantum Entanglement in Time*. arXiv preprint: quant-ph/0402127. Accessed time: 2025/3/2.

Brun T.A., and Halliwell J.J., 1996, Decoherence of hydrodynamic histories: A simple spin model, *Physics Review D*, 54(4), 2899–2912.

Bub J., 2022, Hidden variables, in Olival Freire J.R., et al., eds., *The Oxford Handbook of the History of Quantum Interpretations*, pp. 957–985, Oxford University Press.

Burch A., 2004, Histories electromagnetism, *Journal of Mathematical Physics*, 45(6), 2153–2170.

Cabello A., 2017, Interpretations of quantum theory: A map of madness, in Lombardi O., Fortin S., Holik F. and López C., eds., *What Is Quantum Information?*, pp. 138–143, Cambridge University Press.

Cabello A., Estebaranz J.M. and García-Alcaine G., 1996, Bell-Kochen-Specker theorem: A proof with 18 vectors, *Physics Letters A*, 212, 183–187.

Chen E.K., 2019, Realism about the wave function, *Philosophy Compass*, 14(7), e12611. https://compass.onlinelibrary.wiley.com/doi/10.1111/phc3.12611

Chen E.K., 2021, Quantum mechanics in a time-asymmetric universe: On the nature of the initial quantum state, *The British Journal for the Philosophy of Science*, 72(4), 1155–1183.

Cheung P.T., 2020, *Decoherent Histories Approach: A Quantum Description of Closed Systems*, Master Thesis of Theoretical Physics Group of Imperial College London.

Christodoulakis T. and Wallden P., 2011, The problem of time in quantum cosmology: A decoherent histories view, *Journal of Physics: Conference Series*, 283(1), 012041. IOP Publishing.

Ciaglia F.M., Ibort A. and Marmo G., 2019a, Differential geometry of quantum states, observables and evolution, in Ballico E., Bernardi A., Carusotto I., Mazzucchi S. and Moretti V., eds., *Quantum Physics and Geometry*, pp. 153–172, Springer.

Ciaglia F.M., Ibort A. and Marmo G., 2019b, Schwinger's picture of quantum mechanics I: Groupoids, *International Journal of Geometric Methods in Modern Physics*, 16(8), 1950119.

Ciaglia F.M., Di Cosmo F., Ibort A. and Marmo G., 2020, Evolution of classical and quantum states in the groupoid picture of quantum mechanics, *Entropy*, 22(11), e22111292.

Clarke C.J.S., 2001, The histories interpretation: Stability instead of consistency? *Foundations of Physics Letters*, 14, 179–186.

Cohen H.F., 2010, *How Modern Science Came into the World: Four Civilizations, One 17th-Century Breakthrough*, Amsterdam University Press.

Cotler J. and Wilczek F., 2015, *Bell Tests for Histories*. https://arxiv.org/abs/1503.06458v1. Accessed time: 2025/3/2.

Cotler J. and Wilczek F., 2016, Entangled histories, *Physica Scripta*, T168, 014004. https://dx.doi.org/10.1088/0031-8949/2016/T168/014004

Cotler J. and Wilczek F., 2017, *Temporal Observables and Entangled Histories*. https://arxiv.org/abs/1702.05838v1. Accessed time: 2025/3/2.

Cotler J., et al., 2017, Experimental test of entangled histories, *Annals of Physics*, 387, 334–347. https://doi.org/10.1016/j.aop.2017.09.004

Cotler J., et al., 2018, Superdensity operators for spacetime quantum mechanics, *Journal of High Energy Physics*, 2018(9), 1–57. https://doi.org/10.1007/JHEP09(2018)093

Cotler J., et al., 2019, Quantum causal influence, *Journal of High Energy Physics*, 2019(7), 1–67. https://doi.org/10.1007/JHEP07(2019)042

Craig D. and Singh P., 2011, Consistent histories in quantum cosmology, *Foundations of Physics*, 41(3), 371–379. https://doi.org/10.1007/s10701-010-9422-6

Curiel E., 2009, General relativity needs no interpretation, *Philosophy of Science*, 76(1), 44–72. https://doi.org/10.1086/599277

David F., 2015, *The Formalisms of Quantum Mechanics*, Springer International Publishing Switzerland.

de Broglie L., 1927, La mécanique ondulatoire et la structure atomique de la matière et du rayonnement, *Journal de Physique et le Radium*, 8(5), 225–241.

Desmet R. and Irvine A.D., 2022, Alfred North Whitehead, in Zalta E.N. and Nodelman U., eds., *The Stanford Encyclopedia of Philosophy* (Winter 2022 Edition), https://plato.stanford.edu/archives/win2022/entries/whitehead/. Accessed time: 2025/3/2.

DeWitt B.S. and Graham N., 1973, *The Many Worlds Interpretation of Quantum Mechanics*, Princeton University Press.

Dewar N., 2021, *Theorem of the Week: The Łos-Tarski Preservation Theorem,* https://web.archive.org/web/20220702172155/https://neildewar.wordpress.com/2021/03/03/therem-of-the-week-the-los-tarski-preservation-theorem/. Accessed time: 2025/3/2.

Dieks D., 2017, Underdetermination, realism and objectivity in quantum mechanics, in Agazzi E., ed., *Varieties of Scientific Realism: Objectivity and Truth in Science*, pp. 295–314, Springer.

Diósi L., 1994, *Critique of Weakly Decohering and Linearly Positive Histories*, preprint, https://arxiv.org/abs/gr-qc/9409017. Accessed time: 2025/3/2.

Dirac P.A.M., 1930, *The Principles of Quantum Mechanics*, Clarendon Press.

Dirac P.A.M., 1933, The Lagrangian in quantum mechanics, *Physikalische Zeitschrift der Sowjetunion*, 3(1), 64–72.

Döring A., Eva B. and Ozawa M., 2021, A bridge between Q-worlds, *The Review of Symbolic Logic*, 14(2). https://doi.org/10.1017/S1755020319000492

Döring A. and Isham C., 2010, What is a thing? Topos theory in the foundations of physics, in Abramsky S., Tzevelekos N. and Coecke Bob, eds., *New Structures for Physics*, pp. 753–937. Springer.

Dowker F. and Kent, A., 1996, On the consistent histories approach to quantum mechanics, *Journal of Statistical Physics*, 82(5–6), 1575–1646.

Dowker F., Johnston S. and Sorkin R.D., 2010, Hilbert spaces from path integrals. *Journal of Physics A: Mathematical and Theoretical*, 43(27), 275302. https://iopscience.iop.org/article/10.1088/1751-8113/43/27/275302/meta

Dowker F. and Sorkin R.D., 2023, An intrinsic causality principle in histories-based quantum theory: A proposal. *Journal of Physics A: Mathematical and Theoretical*, 56(48), 485304. https://doi.org/10.1088/1751-8121/ad0347.

Dowker F. and Wilkes H., 2022, An argument for strong positivity of the decoherence functional in the path integral approach to the foundations of quantum theory, *AVS Quantum Science*, 4(1), 012601.

Drummond B., 2019, Understanding quantum mechanics: A review and synthesis in precise language, *Open Physics*, 17, 390–437.

Dürr D., Goldstein S. and Zanghì N., 1996, Bohmian mechanics and the meaning of the wave function, in Shimony A., Cohen R.S., Horne M. and Stachel J.J., eds., *Experimental Metaphysics: Quantum Mechanical Studies in Honor of Abner Shimony, 1*, pp. 25–38, Kluwer Academic Publishers.

Dürr D., Goldstein S. and Zanghì N., 2012, *Quantum Physics Without Quantum Philosophy*, Springer-Verlag.

Epperson M., 2004, *Quantum Mechanics and the Philosophy of Alfred North Whitehead*, Fordham University Press.

Epperson M. and Zafiris, E., 2013, *Foundations of Relational Realism: A Topological Approach to Quantum Mechanics and the Philosophy of Nature,* Lexington Books.

Feynman R.P., 1948, Space-time approach to non-relativistic quantum mechanics, *Review of Modern Physics*, 20, 367–387.

Flori C., 2010, A topos formulation of history quantum theory, *Journal of Mathematical Physics*, 51(5), 053527.

Flori C., 2013, *A First Course in Topos Quantum Theory*, Springer International Publishing.

Flori C., 2018, *A Second Course in Topos Quantum Theory*, Springer International Publishing.

Fröhlich J. and Schubnel B., 2013, *Quantum Probability Theory and the Foundations of Quantum Mechanics*, https://arxiv.org/abs/1310.1484. Accessed time: 2025/3/2.

Fröhlich J., 2020, A brief review of the "ETH-Approach to Quantum Mechanics", in Anantharaman N., et al., eds., *Frontiers in Analysis and Probability*, pp. 21–46, Springer International Publishing.

Fröhlich J., 2021, Relativistic quantum theory, in Allori V., et al., eds., *Do Wave Functions Jump?*, pp. 237–258, Springer International Publishing.

Fröhlich J., et al., 2024, A tentative completion of quantum mechanics, in Bassi A., et al., eds., *Physics and the Nature of Reality*, pp. 151–165, Springer International Publishing.

Gao S., 2017, *The Meaning of the Wave Function: In Search of the Ontology of Quantum Mechanics*, Cambridge University Press.

Gell-Mann M. and Hartle J.B., 1990, Quantum mechanics in the light of quantum cosmology, in Zurek W.H., ed., *Complexity, Entropy and the Physics of Information*, pp. 425–458, Addison-Wesley.

Gell-Mann, M., & Hartle, J. B., 1993, Classical equations for quantum systems, *Physical Review D*, 47(8), 3345–3382.

Gell-Mann M. and Hartle J.B., 1994, Time symmetry and asymmetry in quantum mechanics and quantum cosmology, in Halliwell J.J., Perez-Mercader J. and Zurek W.H., eds., *Physical Origins of Time Asymmetry*, pp. 311–345. Cambridge University Press.

Gell-Mann M. and Hartle J.B., 2014, Adaptive coarse graining, environment, strong decoherence, and quasiclassical realms, *Physical Review A*, 89(5), 052125. https://dx.doi.org/10.1103/PhysRevA.89.052125

Ghazi-Tabatabai Y., 2009, *Quantum Measure Theory: A New Interpretation*, Doctoral Thesis, Imperial College, University of London.

Ghazi-Tabatabai Y., 2013, *Coevents as Beables*, https://arxiv.org/abs/1303.4729v1. Accessed time: 2025/3/2.

Ghirardi G.C., Rimini A. and Weber T., 1986, Unified dynamics for microscopic and macroscopic systems, *Physical review D*, 34(2), 470–491.

Giachetta G., Mangiarotti L. and Sardanashvily, G.A., 2010, *Geometric Formulation of Classical and Quantum Mechanics*, World Scientific Press.

Goldstein S. and Page D.N., 1995, Linearly positive histories: probabilities for a robust family of sequences of quantum events, *Physical Review Letters*, 74(19), 3715–3719.

Goldstein S. and Teufel S., 2001, Quantum spacetime without observers: Ontological clarity and the conceptual foundations of quantum gravity, in Callender C. and Huggett N., eds., *Physics Meets Philosophy at the Planck Scale: Contemporary Theories in Quantum Gravity*, pp. 275–289, Cambridge University Press.

Goldstein S. and Zanghì N., 2013, Reality and the role of the wave function in quantum theory, in Ney A. and Albert D.Z., eds., *The Wave Function: Essays on the Metaphysics of Quantum Mechanics*, pp. 91–119. Oxford University Press.

Greenberger D., Hentschel K. and Weinert F., 2009, *Compendium of Quantum Physics*, Springer.

Griffiths R.B., 1984, Consistent histories and the interpretation of quantum mechanics, *Journal of Statistical Physics*, 36(1–2), 219–272.

Griffiths R.B., 2003, *Consistent Quantum Theory*, Cambridge University Press.

Griffiths R.B., 2013, A consistent quantum ontology, *Studies in History and Philosophy of Science Part B: Studies in History and Philosophy of Modern Physics*, 44(2), 93–114. https://dx.doi.org/10.1016/j.shpsb.2012.12.002

Griffiths R.B., 2014, The new quantum logic, *Foundations of Physics*, 44(6), 610–640. https://dx.doi.org/10.1007/s10701-014-9802-4

Griffiths R.B., 2017, Quantum information: what's it all about? *Entropy*, 19(12), 645. https://dx.doi.org/10.3390/e19120645

Griffiths R.B., 2023. *Consistent Quantum Causes*. arXiv preprint arXiv:2303.13617. Accessed time: 2025/3/2.

Griffiths R.B., 2024, The consistent histories approach to quantum mechanics, in Zalta E.N. and Nodelman U. eds., *The Stanford Encyclopedia of Philosophy* (Summer 2024 Edition), https://plato.stanford.edu/archives/sum2024/entries/qm-consistent-histories/. Accessed time: 2025/3/2.

Griffiths, R.B., and Hartle, J.B., 1998, Comment on "Consistent sets yield contrary inferences in quantum theory". *Physical Review Letters*, 81(9), 1981.

Grygiel W.P., 2007. Consistent quantum histories: Towards a universal language of physics. *Old and New Concepts of Physics*, 4(1), 71–95.

Gudder S., 2011, Quantum measures and the coevent interpretation, *Reports on Mathematical Physics*, 67(1), 137–156.

Guo X.-K., 2020, Tensor networks for quantum causal histories, *Journal of Physics A: Mathematical and Theoretical*, 53, 125301. https://doi.org/10.1088/1751-8121/ab757e

Haag R., 2015, Faces of quantum physics, in Blanchard P. and Fröhlich J. eds., *The Message of Quantum Science: Attempts Towards a Synthesis*, pp. 219–234. Springer.

Haag R., 2019. On quantum theory. *International Journal of Quantum Information*, 17(04), 1950037. https://dx.doi.org/10.1142/S0219749919500370

Hall N., 2015, Humean reductionism about laws of nature, in Barry L. and Jonathan S., eds., *A companion to David Lewis*, pp. 262–277, Wiley-Blackwell John Wiley & Sons.

Halliwell J.J., 2000, The emergence of hydrodynamic equations from quantum theory: A decoherent histories analysis, *International Journal of Theoretical Physics*, 39(7), 1767–1777.

Halliwell J.J., 2003, Decoherence of histories and hydrodynamic equations for a linear oscillator chain, *Physical Review D*, 68(2), 025018.
Halliwell J.J., 2005, How the quantum universe became classical? *Contemporary Physics*, 46(2), 93–104.
Halliwell J.J., 2009a, Partial decoherence of histories and the Diósi test, *Quantum Information Processing*, 8(6), 479–491. https://doi.org/10.1007/s11128-009-0134-9
Halliwell J.J., 2009b, Probabilities in quantum cosmological models: A decoherent histories analysis using a complex potential, *Physical Review D*, 80(12), 124032.
Halliwell, J.J., 2017, Incompatible multiple consistent sets of histories and measures of quantumness, *Physical Review A*, 96(1), 012123. https://doi.org/10.1103/PhysRevA.96.012123
Halliwell J.J. and Yearsley J.M., 2009, Arrival times, complex potentials, and decoherent histories, *Physics Review A*, 79, 062101. https://dx.doi.org/10.1103/PhysRevA.79.062101
Halliwell J.J. and Zafiris E., 1998, Decoherent histories approach to the arrival time problem, *Physical Review D*, 57(6), 3351–3364.
Halvorson H., 2019, To be a realist about quantum theory, in Fortin, S., Holik, F., Lombardi, O. and López, C., eds., *Quantum Worlds: Perspectives on the Ontology of Quantum Mechanics*, pp. 133–163, Cambridge University Press.
Hao L.X. 郝刘祥, 2014, The mathematization of nature: An inquiry into a historiographical issue of the scientific revolution (in Chinese) 自然的数学化——关于科学革命编史纲领的探讨, *Science & Culture Review*科学文化评论, 11(5), 5–23.
Hao L.X. 郝刘祥, 2021a, Philosophy and physics meet in quantum world (in Chinese) 哲学与物理学相遇在量子世界, *Bulletin of Chinese Academy of Sciences*中国科学院院刊, 36(1), 28–36.
Hao L.X. 郝刘祥, 2021b, Representation, intervention and reality: The structural process of scientific revolutions (in Chinese) 表征 、 介入与实在——科学革命的结构化进程, *Journal of Dialectics of Nature*自然辩证法通讯, 43(6), 24–31.
Hao L.X. 郝刘祥, 2024, The dynamical mode of scientific progress (in Chinese) 科学进步的动力学模式, *Science & Technology Review*科技导报, 42(10), 27–39.
Hartle, J.B., 1993, Spacetime Quantum Mechanics and the Quantum Mechanics of Spacetime, *Vistas in Astronomy*, 37, 569–583.
Hartle J.B., 2004, Bohmian histories and decoherent histories, *Physical Review A*, 69(4), 042111.
Hartle J.B., 2005, What connects different interpretations of quantum mechanics? in Elitzur A.C., Dolev S. and Kolenda N., eds., *Quo Vadis Quantum Mechanics?* pp. 73–82. Springer.
Hartle J.B., 2007, Quantum physics and human language, *Journal of Physics A: Mathematical and Theoretical*, 40(12), 3101–3121.
Hartle J.B., 2010, Quasiclassical realms, in Saunders S., Barrett J., Kent A. and Wallace D., eds., *Many Worlds: Everett, Quantum Theory, and Reality*, pp. 73–98. Oxford University Press.
Hartle J.B., 2011, The quasiclassical realms of this quantum universe, *Foundations of Physics*, 41(6), 982–1006.
Hartle J.B., 2016, *Why Our Universe Is Comprehensible?* preprint, https://arxiv.org/abs/1612.01952, reprinted in Hartle J.B., *The Quantum Universe: Essays on Quantum Mechanics, Quantum Cosmology, and Physics in General*, 2021, World Scientific Publishing.
Hartle J.B., 2021, *The Quantum Universe: Essays on Quantum Mechanics, Quantum Cosmology, and Physics in General*, World Scientific Publishing.
Hemmo M., 1996, *Quantum Mechanics Without Collapse: Modal Interpretations, Histories and Many Worlds*, Doctoral Dissertation, University of Cambridge.

Hemmo M., 1998, Quantum histories in the modal interpretation, in Dieks D. and Vermaas P.E., eds., *The Modal Interpretation of Quantum Mechanics*, pp. 253–277, Springer.

Hemmo M. and Shenker O., 2020, Why the many-worlds interpretation of quantum mechanics needs more than Hilbert space structure? in Peels R., et al., eds., *Scientific Challenges to Common Sense Philosophy*, pp. 61–70, Routledge.

Hicks M.T., 2018, Dynamic humeanism, *The British Journal for the Philosophy of Science*, 69 (4), 983–1007. https://doi.org/10.1093/bjps/axx006

Howard D., 2004, Who invented the "Copenhagen Interpretation"? A study in mythology, *Philosophy of Science*, 71(5), 669–682.

Isham C.J., 1993, Canonical quantum gravity and the problem of time, in Ibort, L.A. and Rodríguez, M.A., eds., *Integrable Systems, Quantum Groups, and Quantum Field Theories*, pp. 157–287, Springer.

Isham C.J., Linden N., Savvidou K. and Schreckenberg S., 1998, Continuous time and consistent histories, *Journal of Mathematical Physics*, 39(4), 1818–1834.

Jammer M., 1989, *The Conceptual Development of Quantum Mechanics*, Tomash Publishers, American Institute of Physics.

Joos E. and Zeh H.D., 1985, The emergence of classical properties through interaction with the environment, *Zeitschrift für Physik B: Condensed Matter*, 59(2), 223–243.

Kent A., 1996a, Quasiclassical dynamics in a closed quantum system, *Physical Review A*, 54(6), 4670–4675.

Kent A., 1996b, Remarks on consistent histories and Bohmian mechanics, in Cushing, J.T., Fine, A. and Goldstein, S., eds., *Bohmian Mechanics and Quantum Theory: An Appraisal*, pp. 343–352, Springer.

Kent A., 1997, Consistent sets yield contrary inferences in quantum theory, *Physical Review Letters*, 78(15), 2874–2877.

Kent A., 1998, Quantum histories, *Physica Scripta*, T76, 78–84.

Kochen, S. and Specker, E.P., 1967, The Problem of Hidden Variables in Quantum Mechanics. *Journal of Mathematics and Mechanics*, 17, 59–87

Kronz F. and Tracy L., 2025, Quantum theory and mathematical rigor, in Zalta E.N. and Nodelman U. eds., *The Stanford Encyclopedia of Philosophy* (Spring 2025 Edition), https://plato.stanford.edu/archives/spr2025/entries/qt-nvd/. Accessed time: 2025/3/2.

Kuhlmann, M., 2023, Quantum field theory, in Zalta E.N. and Nodelman U., eds., *The Stanford Encyclopedia of Philosophy* (Summer 2023 Edition), https://plato.stanford.edu/archives/sum2023/entries/quantum-field-theory/. Accessed time: 2025/3/2.

Laloë F., 2019, *Do We Really Understand Quantum Mechanics?* (second edition), Cambridge University Press.

Laloë F., 2022, Quantum mechanics is routinely used in laboratories with great success, but no consensus on its interpretation has emerged, in Olival Freire J.R., et al., eds., *The Oxford Handbook of the History of Quantum Interpretations,* pp. 7–51, Oxford University Press.

Landsman K., 2017, *Foundations of Quantum Theory: From Classical Concepts to Operator Algebras*, Springer.

Laura R. and Vanni, L., 2009, Time translation of quantum properties, *Foundations of Physics*, 39. https://doi.org/10.1007/s10701-008-9268-3

Leifer M., 2005, *Quantum Logic, in Interpretation of Quantum Mechanics: Current Status and Future Directions,* Lecture 22, Perimeter Institute for Theoretical Physics, https://www.slideshare.net/mleifer/quantum-logic.

Li S.W., Cai C.Y., Liu X.F. and Sun C.P., 2018, Objectivity in quantum measurement, *Foundations of Physics*, 48, 654–667. https://doi.org/10.1007/s10701-018-0169-9

Loewer B., 2001, Determinism and chance, *Studies in History and Philosophy of Science Part B: Studies in History and Philosophy of Modern Physics*, 32(4), 609–620. https://doi.org/10.1016/S1355-2198(01)00028-4

Losada M., 2018, Contrary quantum histories and contrary inferences, *Physica A: tatistical Mechanics and Its Applications*, 503, 379–389. https://doi.org/10.1016/j.physa.2018.02.214

Losada, M. and Laura, R., 2014a, Quantum histories without contrary inferences. Annals of Physics, 351, 418–425.

Losada M. and Laura R., 2014b, Generalized contexts and consistent histories in quantum mechanics, *Annals of Physics*, 344, 263–274. https://dx.doi.org/10.1016/j.aop.2014.03.001

Losada M. and Lombardi O., 2018, Histories in quantum mechanics: Distinguishing between formalism and interpretation, *European Journal for Philosophy of Science*, 8(3). https://doi.org/10.1007/s13194-017-0197-x

Losada M., Laura R. and Lombardi O., 2019, Frauchiger-Renner argument and quantum histories. *Physical Review A*, 100(5), 052114. https://doi.org/10.1103/PhysRevA.100.052114

Mandolesi A.L., 2017, *Non-Probabilistic Decoherent Histories and Causal Histories in Everettian Quantum Mechanics*, preprint, https://arxiv.org/abs/1704.01173. Accessed time: 2025/3/2.

Madelung E., 1927, Quantentheorie in hydrodynamischer form, *Zeitschrift für Physik*, 40, 322–326.

Marlow T., 2006, A Bayesian account of quantum histories, *Annals of Physics*, 321(5), 1103–1125. https://doi.org/10.1016/j.aop.2005.09.006

McDonnell J.F., 2019. Quantum monadology. *Idealistic Studies*, 47(3), 219–235. https://doi.org/10.5840/idstudies201912587

McKenzie K., 2021. Science-guided metaphysics, in Bliss, R. and Miller, J.T.M., eds., *The Routledge Handbook of Metametaphysics*, pp. 435–446. Routledge.

Mermin N.D., 2012, Commentary: Quantum mechanics: Fixing the shifty split, *Physics Today*, 65(7). https://doi.org/10.1063/PT.3.1618

Muller, F. A., 2015. Circumveiloped by Obscuritads: The nature of interpretation in quantum mechanics, hermeneutic circles and physical reality, with cameos of James Joyce and Jacques Derrida, in Béziau J.-Y., Krause D., and Arenhart J.R.B., eds., *Conceptual clarifications. Tributes to Patrick Suppes (1922–2014)*, pp. 107–136, College Publications.

Milz S., et al., 2021, Genuine multipartite entanglement in time, *SciPost Physics*, 10, 141. https://scipost.org/SciPostPhys.10.6.141

Nowakowski M., 2017, Quantum entanglement in time, *AIP Conference Proceedings*, 1841, 020007. https://dx.doi.org/10.1063/1.4982771.

Okon E. and Sudarsky D., 2014a, On the consistency of the consistent histories approach to quantum mechanics, *Foundations of Physics*, 44(1), 19–33. https://dx.doi.org/10.1007/s10701-013-9760-2.

Okon E. and Sudarsky D., 2014b, Measurements according to consistent histories, *Studies in History and Philosophy of Science Part B: Studies in History and Philosophy of Modern Physics*, 48. https://dx.doi.org/10.1016/j.shpsb.2014.08.011

Okon E. and Sudarsky D., 2015, The consistent histories formalism and the measurement problem, *Studies in History and Philosophy of Science Part B: Studies in History and Philosophy of Modern Physics*, 52, 217–222. https://dx.doi.org/10.1016/j.shpsb.2015.07.008

Olival Freire J.R., et al., eds., 2022, *The Oxford Handbook of the History of Quantum Interpretations*, Oxford University Press.

Omnès R., 1988, Logical reformulation of quantum mechanics. I: Foundations, *Journal of Statistical Physics*, 53(3–4), 893–932.

Omnès R., 1994, *The Interpretation of Quantum Mechanics*, Princeton University Press.

Omnès R., 1999, *Understanding Quantum Mechanics*, Princeton University Press.

Omnès R., 2004, The language of interpretation in quantum physics and its logic, in Weingartner, P., ed., *Alternative Logics. Do Sciences Need Them?*, pp. 249–256, Springer.

Ozawa M., 2016, Quantum set theory extending the standard probabilistic interpretation of quantum theory, *New Generation Computing*, 34(1–2), 125–152.

Ozawa M., 2021, From boolean valued analysis to quantum set theory: Mathematical worldview of Gaisi Takeuti, *Mathematics*, 9, 937. https://doi.org/10.3390/math9040397

Petersen A., 1963, The philosophy of Niels Bohr, *Bulletin of the Atomic Scientists*, 19(7), 8–14. https://doi.org/10.1080/00963402.1963.11454520

Priebe R., 2012, *The Regular Histories Formulation of Quantum Theory*. Doctoral Dissertation, University of Oxford.

Riedel C.J., 2014, *Consistency Conditions in Consistent Histories*, https://blog.jessriedel.com/2014/03/18/consistency-conditions-in-consistent-histories. Accessed time: 2025/3/2.

Riedel C.J., Zurek W.H. and Zwolak M., 2016, Objective past of a quantum universe: Redundant records of consistent histories, *Physical Review A*, 93(3), 032126. https://dx.doi.org/10.1103/PhysRevA.93.032126

Roberts B.W. and Teh N.J., 2016, *Kähler Representation Theory*, preprint, https://arxiv.org/abs/1602.05979. Accessed time: 2025/3/2.

Robson C., 2024, Relational quantum mechanics and contextuality, *Foundations of Physics*, 54. https://doi.org/10.1007/s10701-024-00788-0

Rocha G.R., Rickles D. and Boge F.J., 2022, A brief historical perspective on the consistent histories interpretation of quantum mechanics, in Olival Freire J.R., et al., eds., *The Oxford Handbook of the History of Quantum Interpretations*, pp. 1175–1196. Oxford University Press.

Rovelli C., 1996, Relational quantum mechanics, *International Journal of Theoretical Physics*, 35(8), 1637–1678.

Rovelli C., 2017, *Reality Is Not What It Seems: The Journey to Quantum Gravity*, Penguin Books.

Rovelli C., 2022, Quantum ontology without wave function, in Allori V., ed., *Quantum Mechanics and Fundamentality*, Synthese Library 460, pp. 157–162, Springer.

Rovelli C., 2025, Relational quantum mechanics, in Zalta E.N. and Nodelman U., eds., *The Stanford Encyclopedia of Philosophy* (Spring 2025 Edition), https://plato.stanford.edu/archives/spr2025/entries/qm-relational/. Accessed time: 2025/3/2.

Ruetsche L., 2011, *Interpreting Quantum Theories*, Oxford University Press.

Ruetsche, L., 2024, The miracles argument meets quantum mechanics, *Theoria: An International Journal for Theory, History and Foundations of Science*, 39(2), 245–261. https://doi.org/10.1387/theoria.24976

Ruyant Q., 2023, Consistent histories through pragmatist lenses, *Studies in History and Philosophy of Science*, 98, 40–48. https://doi.org/10.1016/j.shpsa.2023.01.001

Salimkhani K. and Rolffs M., 2022, *On Progress in Metaphysics*. https://philsci-archive.pitt.edu/21178/. Accessed time: 2025/3/2.

Savvidou K., 1999a, *Continuous Time and Consistent Histories*, Doctoral Dissertation, Imperial College of Science, Technology and Medicine.

Savvidou K., 1999b, The action operator for continuous-time histories, *Journal of Mathematical Physics*, 40(11), 5657–5674.

Savvidou N., 2001, General relativity histories theory: Spacetime diffeomorphisms and the Dirac algebra of constraints, *Classical and Quantum Gravity*, 18(17), 3611–3628.

Schlosshauer M., 2004, Decoherence, the measurement problem, and interpretations of quantum mechanics. *Reviews of Modern physics*, 76(4), 1267.

Schrödinger E., 1927a, *On the Relation between the Quantum Mechanics of Heisenberg, Born, and Jordan and that of Schrödinger*, reprinted in Schrödinger, E., *Collected Papers on Wave Mechanics*, translated by Shearer J.F., Chelsea.

Schrödinger E., 1927b, *Quantisation as a Problem of Proper Values, Part I-II*, reprinted in Schrödinger E., *Collected Papers on Wave Mechanics*, translated by Shearer J.F., Chelsea.

Schrödinger E., 1927c, *Collected Papers on Wave Mechanics*, translated by Shearer J.F., Chelsea.

Schroeren, D.P., 2013, Decoherent histories of spin networks. *Foundations of Physics*, 43(3), 310–328.

Sinha U., Couteau C., Jennewein T., Laflamme R. and Weihs G., 2010, Ruling out multi-order interference in quantum mechanics. *Science,* 329(5990), 418–421. https://doi.org/10.1126/science.1190545

Smolin L., 2021, *Views, Variety and Quantum Mechanics*, preprint, https://arxiv.org/abs/2105.03539. Accessed time: 2025/3/2.

Solé A. and Hoefer C., 2019, The nomological interpretation of the wave function, in Cordero A., ed., *Philosophers Look at Quantum Mechanics*, pp. 119–140, Springer.

Sorkin R.D., 1994, Quantum mechanics as quantum measure theory, *Modern Physics Letters A*, 9(33), 3119–3127.

Sorkin R.D., 1997, Quantum measure theory and its interpretation, in Feng D.H. and Hu B.-L., eds., *Quantum Classical Correspondence: Proceedings of the 4th Drexel Symposium on Quantum Nonintegrability*, Philadelphia, September 8-11, 1994, pp. 229–251, International Press, Cambridge.

Sorkin R.D., 2007, An exercise in "anhomomorphic logic", *Journal of Physics: Conference Series*, 67(1), 012018. IOP Publishing.

Sorkin R.D., 2012, Logic is to the quantum as geometry is to gravity, in Murugan J., Weltman A. and Ellis G.F.R., eds., *Foundations of Space and Time: Reflections on Quantum Gravity,* pp. 363–384. Cambridge University Press.

Stapp H.P., 2004, *Mind, Matter, and Quantum Mechanics*, Springer.

Stapp H.P., 2012, Quantum Locality? *Foundations of Physics*, 42, 647–655. https://doi.org/10.1007/s10701-012-9632-1

Stoeger W., 2001, Epistemological and ontological issues arising from quantum theory, in Russell, R.J. et al., eds., *Quantum Mechanics: Scientific Perspectives on Divine Action*, pp. 81–98. Vatican Observatory Publications.

Sun C.-P. 孙昌璞, 2017, On interpretations of quantum mechanics (in Chinese) 量子力学诠释问题. *Physics* 物理, 46(8), 481–498.

Sun C.-P. 孙昌璞, 2024, The methodology of physics in pursuing the beauty of truth: From the falsifiability of science to Occam's razor principle (in Chinese) 物理学求美至真的方法论——从科学可证伪性到奥卡姆剃刀原则. *Science & Technology Review*科技导报, 42(10), 46–54.

Takeuti G., 1981, Quantum set theory, in Beltrametti E.G. and Van Fraassen B.C., eds., *Current Issues in Quantum Logic*, pp. 303–322. Springer.

Toader I.D., 2021, On the categoricity of quantum mechanics, *European Journal for Philosophy of Science*, 11(1), 17. https://doi.org/10.1007/s13194-020-00328-3

Von Neumann J., 1932, *Mathematische Grundlagen der Quantenmechanik*, Springer; English translated by Beyer R.T., *Mathematical Foundations of Quantum Mechanics*, 2018, Princeton University Press.

Wallace D., 2008, Philosophy of quantum mechanics, in Rickles D., ed., *The Ashgate Companion to Contemporary Philosophy of Physics*, Routledge.

Wallace D., 2019, What is orthodox quantum mechanics? in Cordero A., ed., *Philosophers Look at Quantum Mechanics*, Springer.

Wallden P., 2013, The coevent formulation of quantum theory, *Journal of Physics Conference Series*, 442(1). https://doi.org/10.1088/1742-6596/442/1/012044

Whitehead A.N., 1929[1985], *Process and Reality (Gifford Lectures 1927–28)*, Macmillan. Corrected edition, Griffin D.R. and Sherburne D. W., eds., The Free Press, 1985.

Whitehead A. N., 1941, Mathematics and the good, in Schilpp P.A., ed. *The Philosophy of Alfred North Whitehead*, pp. 666–681, Northwestern University.

Wigner E.P., 1963, The problem of measurement, *American Journal of Physics*, 31(A963). Reprinted in Wheeler, J.A. and Zurek, W.H., eds., *Quantum Theory and Measurement*, 1983, Princeton University Press.

Wilkes H., 2019, *The Form and Interpretation of the Decoherence Functional*, Doctoral Dissertation, Imperial College London.

Yearsley J.M., 2010, A review of the decoherent histories approach to the arrival time problem in quantum theory, *Journal of Physics Conference Series*, 306(1), 012056. https://doi.org/10.1088/1742-6596/306/1/012056

Zeh H.D., 1970, On the interpretation of measurement in quantum theory, *Foundations of Physics*, 1(1), 69–76. https://link.springer.com/article/10.1007/BF00708656

Zhang T., Dahlsten O. and Vedral V., 2020, *Quantum Correlations in Time*. arXiv preprint, arXiv:2002.10448. Accessed time: 2025/3/2.

Zurek W.H., 1982, Environment-induced superselection rules, *Physical Review D*, 26(8), 1862–1880. https://link.aps.org/doi/10.1103/PhysRevD.26.1862

Zurek W.H., 1991, Decoherence and the transition from quantum to classical, *Physics Today*, 44(10), 36–44.

Afterwords

My earliest interest in the subject of this book – quantum histories interpretation – originated from Professor Liuxiang Hao's advice on my doctoral thesis topic in 2017. Since I began to pay attention to all the amazing and thought-provoking works of Griffiths, Omnès, Hartle, Gell-Mann, Halliwell, Sorkin, et al., I have gained many new comprehensions of the entire landscape of quantum mechanics interpretation and the resolution of quantum paradoxes. Approaching quantum physics and even quantum gravity theory from this perspective on historicized ontology is a very novel but also challenging initiative. And the numerous new philosophical concepts and ideas contained therein will bring unexpected interesting insights and inspirations to the philosophy of physics and scientific metaphysics. This book is an immature work that merely serves as a starting point. I hope to further extend the discussion of the many issues touched upon in Chapter 9 in the future to advance these studies, and I also hope that the reader will be interested in this field, participate in it, and offer their unique insights.

I would like to express my deep gratitude to Professor Liuxiang Hao, the School of Humanities at the University of Chinese Academy of Sciences, the Institute of Philosophy of the Chinese Academy of Social Sciences, and Routledge Publishing Group for their support for my writing.

Index

FSC
www.fsc.org
MIX
Papier | Fördert gute Waldnutzung
FSC® C083411

Kolibri 360 GmbH
Ferdinand-Jühlke-Straße 7
99095 Erfurt, Deutschland
produktsicherheit@kolibri360.de